CCPS Middle East Process Safety Conference 2017

Sanabis, Bahrain
9 - 11 October 2017

ISBN: 978-1-5108-5221-1

Printed from e-media with permission by:

Curran Associates, Inc.
57 Morehouse Lane
Red Hook, NY 12571

Some format issues inherent in the e-media version may also appear in this print version.

Printed by Curran Associates, Inc. (2018)

For permission requests, please contact AIChE
at the address below.

AIChE
120 Wall Street, FL 23
New York, NY 10005-4020

Phone: (800) 242-4363
Fax: (203) 775-5177

www.aiche.org

Additional copies of this publication are available from:

Curran Associates, Inc.
57 Morehouse Lane
Red Hook, NY 12571 USA
Phone: 845-758-0400
Fax: 845-758-2633
Email: curran@proceedings.com
Web: www.proceedings.com

TABLE OF CONTENTS

PROCESS SAFETY IMPLEMENTATION I

POSTER SESSION

ABS Group

MIDDLE EAST PROCESS SAFETY CONFERENCE | 9 - 11 OCTOBER 2017 | Bahrain International Exhibition and Convention Centre

CCPS MIDDLE EAST PROCESS SAFETY CONFERENCE

Ten Years of Living Risk Based Process Safety

Steve Arendt, PE
VP-Global Oil, Gas & Chemicals
+1-281-673-2914
sarendt@absconsulting.com
www.abs-group.com

Steve Arendt, P.E., Vice President - Global Oil, Gas, and Chemicals

- 40 years in process safety and risk assessment
- Conducted 100s of PSM system design/implementations, audits, incident investigations, HAZID/HAZOP/LOPA/QRAs, best practice reviews, and safety culture evaluations
- ABS Group manager for the BP-Baker Panel PSM reviews
- 80+ articles and books on PSM and risk management, including CCPS Guidelines on
  - **Risk Based Process Safety**
  - **Implementing PSM, 2nd edition**
  - **Management of Change**
  - **Guidelines for Hazard Evaluation Procedures, 2nd Edition**
- CCPS Fellow and recipient of MKOPSC Merit Award

ABS Group

MIDDLE EAST PROCESS SAFETY CONFERENCE | 9 - 11 OCTOBER 2017

Discussion Outline

- RBPS background
- Essence of RBPS
- Ways RBPS has been used
- Misconceptions, misuse and possible future improvements
- Closing remarks

ABS Group

RBPS Background

- CCPS published its original 12 PSM elements in 1989 and followed it with 3 other management system books thru 1994
- A lot of experience and lessons were learned from then thru 2005; CCPS wanted to update its PSM framework to be useful to industry as a thought and action leader for the next 20 years
- RBPS came about for two reasons:
  - **To be able to generate better results with fewer resources**
  - **To provide a "target" and an approach for companies of all needs to implement, correct, and improve PSM systems**

ABS Group

RBPS Development Process

- 20 subcommittee members representing the process industry
- 16 peer reviewers
- 2+ year effort that:
  - Surveyed PSM/ESH systems from around the world to identify good features
  - Benchmarked with over 60 companies in three workshops to compile leading practices, improvement idea, and metrics
- 760-page guideline written by a team from ABS Consulting

ABS Group

MIDDLE EAST PROCESS SAFETY CONFERENCE | 9 - 11 OCTOBER 2017

5

RBPS Subcommittee and Co-Authors

Subcommittee

Jack McCavit	*CCPS Emeritus, Committee Chair*
Bob G. Perry	*Center for Chemical Process Safety*
Don Abrahamson	*Celanese Chemical*
Steve Arendt	*ABS Consulting*
Tim Blackford	*Chevron Energy Technology Company*
John Herber	*3M Company*
Dan Isaacson	*The Lubrizol Corporation*
Shakeel Kadri	*Air Products and Chemicals, Inc.*
Greg Keeports	*Rohm and Haas Company*
Jim Klein	*DuPont*
Pete Lodal	*Eastman Chemical Company*
Bill Marshall	*Eli Lilly and Company*
Darren Martin	*Shell Chemical Company*
Neil Maxson	*Bayer Material Science*
Lisa Morrison	*BP*
Karen Tancredi	*DuPont*
Tony Thompson	*Monsanto Company, Retired*
Scott Wallace	*Olin Corporation*
Roy Winkler	*INEOS Olefins and Polymers USA*
Gary York	*Rhodia, Inc., Retired*

Co-Authors

Steve Arendt
Bill Bradshaw, *Project Manager*
Walt Frank
Don Lorenzo
Lee Vanden Heuvel

ABS Group

MIDDLE EAST PROCESS SAFETY CONFERENCE | 9 - 11 OCTOBER 2017

Guidelines for *Risk Based Process Safety*

ABS Group

MIDDLE EAST PROCESS SAFETY CONFERENCE | 9 - 11 OCTOBER 2017

The Essence of Risk Based Process Safety

- Management systems should be the simplest that they can be while still being fit-for-purpose
- Consider the following when determining management system "rigor"
  - **Perception of complexity, hazard, and risk**
  - **Demand for the system results and the resources required to deliver them**
  - **Current company/facility culture**
- To design, correct, and improve process safety management activities
- RBPS elements can be used throughout the company, plant or project life-cycle

ABS Group

MIDDLE EAST PROCESS SAFETY CONFERENCE | 9 - 11 OCTOBER 2017

RBPS System Element Resources

- Element
  - Key Principles
    - Essential Features
      - ➢Possible Work Activities
        - ❖Implementation Examples
          - Performance Improvement Practice
          - Efficiency Improvement Practice
      - ➢List of metrics – LEADING and lagging
      - ➢Management review items

ABS Group

MIDDLE EAST PROCESS SAFETY CONFERENCE 9 - 11 OCTOBER 2017

9

9

RBPS Elements – Relationship to OSHA PSM

RBPS Element	New Element	Expanded Scope	Improved Practices
Process Safety Culture	✓		
Compliance to Standards	✓		
Process Safety Competency	✓		
Workforce Involvement		✓	✓
Stakeholder Outreach	✓		
Process Knowledge Management		✓	✓
Hazard Identification and Risk Analysis		✓	✓
Operating Procedures			✓
Safe Work Practices			✓
Asset Integrity and Reliability		✓	✓
Contractor Management			✓
Training and Performance			✓
Management of Change		✓	✓
Operational Readiness		✓	✓
Conduct of Operations	✓		
Emergency Management			✓
Incident Investigation			✓
Measurement and Metrics	✓		
Auditing			✓
Management Review and Continuous Improvement	✓		

ABS Group

MIDDLE EAST PROCESS SAFETY CONFERENCE 9 - 11 OCTOBER 2017

10

Applications of RBPS during First 10 Years

- Ways that RBPS has been used
- Most frequently adopted new elements/practices
- Most under-used elements and practices
- Uncertainties from companies about how/whether certain PS aspects are addressed
- Ideas on how to improve RBPS framework and approach

ABS Group

MIDDLE EAST PROCESS SAFETY CONFERENCE | 9 - 11 OCTOBER 2017

Ways that RBPS Has Been Used

- Improve/optimize HSE/PS Management Systems
  - Design/implement new system
  - Add new elements/upgrade existing elements with new element activities
  - Fix broken system – metrics, work activities and implementation options
  - Improve existing system – improvement and Management Review ideas
- As primary PSM framework
- As a Process Safety competency training program framework
- As best practice assessment/audit protocol
- As organizational culture assessment basis
- Basis for RCA structure
- Built detailed key process workflow diagrams
- Translated into other languages – Portuguese, Spanish, Chinese
- Other applications? - Benchmarking Risk Based Process Safety Egg (RBPS Temple + 2 Universes) Against "The Art of War"

Most Frequent
Occasional Use
Rare

ABS Group

MIDDLE EAST PROCESS SAFETY CONFERENCE | 9 - 11 OCTOBER 2017

Misconceptions/Uncertainties

- Use of "max" implementation options as basis for audit protocol
  - Elements Possible Work Activities were defined with a range of implementation option from minimal to very detailed
  - Some who developed audit protocol used the most detailed option as the implementation standard for determining nonconformance – not taking into consideration the hazard/risk/complexity, resources and culture needs of the specific company/plant circumstances
- How leadership commitment is addressed in RBPS – part of culture element
- How Human Factors is addressed in RBPS – put into relevant elements, not as a single “cross-cutting” element by itself
- Use of “all/many” element metrics provided in ach element chapter

ABS Group

MIDDLE EAST PROCESS SAFETY CONFERENCE

Underused Elements/Aspects

- Process Knowledge Management – applying for knowledge
- Stakeholder Outreach
- Conduct of Operations (CONOPs)
  - For explicit list is areas where workforce Operational Discipline is key
  - Especially for management operational discipline
- Efficiency improvement ideas contained in each element chapter
- Develop a key process/workflow for each HSE/PSM element using RBPS work activities as design toolbox
  - Good for a specific company or facility
  - Difficult for an industry/organization to if need “spectrum” in broad

ABS Group

MIDDLE EAST PROCESS SAFETY CONFERENCE

Improvement Ideas for the RBPS 2nd Edition

- Update/streamline leading indicator metrics
- Update performance improvement ideas
- Update efficiency improvement ideas
- Address Cybersecurity and resilience somehow – other new elements?
- More explicit use for design organizations and capital project management
- Direct toward use as a means for achieving Operational Excellence
- When will RBPS 2.0 be needed?
  - Probably not for a few years; the current edition is "still being discovered" by certain regions and industries

ABS Group

MIDDLE EAST PROCESS SAFETY CONFERENCE | 9 - 11 OCTOBER 2017

CCPS Guidelines for *Implementing PSM 2nd Ed*

Laundry

- Commit to Process Safety
- Understand Ha zards & Risk
- Manage Risk
- Learn from Experien

Gas Station

- Safety information
- Hazard analysis/JSA
- Procedures
- Training
- nce
- Emergen
- Investigation
- Audits

Low Hazard/Simple

- Ensure management cares/resources
- Ensure employees care
- Demonstrate stakeholder commitment
- Know what you operate
- Reduce or eliminate hazards
- e risk
- Und idual risk
- Know how to operate
- Know how to maintain
- Control changes
- Prepare/respond/manage incidents
- Monitor/act on internal learning
- Monitor/act on external sources

High Hazard/Complex

- Culture
- Standards
- Competency
- Involvement
- Outreach
- Knowledge
- Hazards/Risk
- Procedures
- SW Practices
- Integrity
- actors
- P rformance
- MOC
- Readiness
- Discipline
- Emergencies
- Investigations
- Metrics
- Auditing
- Management Review

RBPS Design Criteria

- Hazard/complexity/risk
- Resources
- Initial culture

Elements

- Key principles
- Essential features
- Work activities
- Implementation options
- Improvement ideas
- Metrics and MR

ABS Group

VISION 20/20 | CCPS

An AIChE Technology Alliance
Center for Chemical Process Safety

- **Industry Tenets**
  - Committed culture
  - Vibrant management systems
  - Disciplined adherence to standards
  - Intentional competency development
  - Enhanced application and sharing of lessons learned

Societal Themes

- Enhanced Stakeholder Knowledge
- Responsible Collaboration
- Harmonization of Standards
- Meticulous Verification

Assessment Tool

Vision 20/20, developed by the Center for Chemical Process Safety (CCPS), looks into the not-too-distant future to describe how great process safety is delivered when it is collectively and fervently supported by industry, regulators, academia, and the community worldwide; driven by the five industry tenets; and enhanced by the four global societal themes.

ABS Group

MIDDLE EAST PROCESS SAFETY CONFERENCE | 9 - 11 OCTOBER 2017

VISION 20/20 | CCPS

An AIChE Technology Alliance
Center for Chemical Process Safety

Implementing the Industry Tenets…

Prepare	Assess	Plan	Perform	*Achieve*	Sustain
Present V20/20 to PSM Colleagues and Management	Complete the V20/20 Assessment Tool	Reinforce and Use Strong Elements as Building Blocks	Implement Action Plans	Complete Action Plans	Verify Management System Improvements
Make V20/20 a Regular Topic at PSM-Related Meetings	Identify Weak and Strong Sub-Elements (<2.5 or >3.5 Respectively)	Identify the Specific Improvements Needed	Monitor Status of Action Plan Implementation	Re-Assess V20/20 Implementation Status with the Assessment Tool	Develop Action Plans for Weak Sub-Elements and Individual Items
"Sprinkle" V20/20 into PSM Conversations	Report Results; Management Commits to Improve	Research Options to Improve (Reference Industry Documents)	Evaluate Effectiveness of Actions	Report & Celebrate Improvements	Implement Action Plans and Monitor Performance
Use V20/20 Logo on Internal Communications	Communicate Results Within Organization	Develop Specific Action Plans to Address Weak Areas	Capture & Communicate Learnings	Identify New Weak Sub-Elements and Weak Individual Items (<2)	Continual Improvement… Continue the Journey!

Today → *2020*

ABS Group

MIDDLE EAST PROCESS SAFETY CONFERENCE | 9 - 11 OCTOBER 2017

RBPS Trivia

Dedication	1
Bosses	2
RBPS Criteria	3
Pillars	4
Authors	5
Contributors	7
Figures	8
Reviewers	16
Elements	20
Subcommittee	20
Tables	26
Total People	48
Acronyms	65
Key Practices	78
Terms	99
References	117
Management Review Items	207
Index Terms	219
Improvement Ideas	306
Metrics	319
Essential Features	319
Work Activities	622
Pages	760
Implementation Options	2,042
Paragraphs	9,017
Lines	29,604
Words	254,233
Characters	1,450,964
Spaces	1,703,022
MSWord File Bytes	5,253,632
Priceless	RBPS source

ABS Group

MIDDLE EAST PROCESS SAFETY CONFERENCE | 9 - 11 OCTOBER 2017

Dedication

Guidelines for Risk Based Process Safety
Dedicated to the memory of

Lester H. Wittenberg
[illegible]

Les was among the first employees of the Center for Chemical Process Safety and was a major contributor to its success. All who knew and worked with Les remember him for his knowledge, his courage, and for his love of life. Although he is gone, his smile is still with us.

ABS Group

MIDDLE EAST PROCESS SAFETY CONFERENCE | 9 - 11 OCTOBER 2017

Closing Comments

- Use RBPS to help "Manage generational change" in Process Safety
- Rediscover use of the RBPS toolbox
- Recommend CCPS implement online RBPS use/needs survey for near future to capture needs and improvement ideas for RBPS 2.0 or ...whatever...

ABS Group

ABS Group

22

Air Products Process Safety Design for World-Scale Air Separation Plants, as Applied On The Jazan Project

Jim Kourpas

Derek Miller

James VanOmmeren

Fiona Kirkham

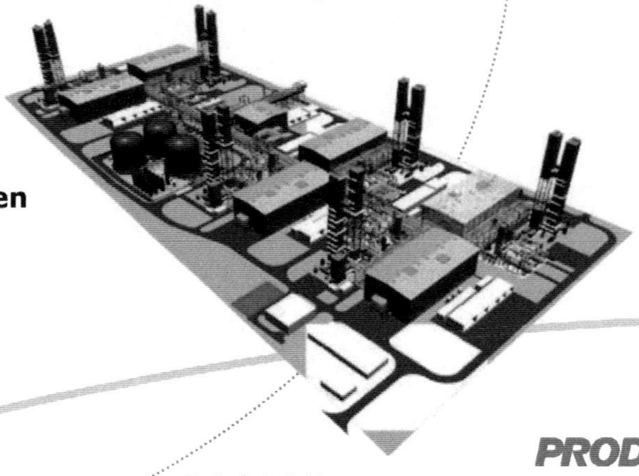

1 Air Products Public

AIR PRODUCTS

Site Overview

Jim Kourpas

VIDEO AND BRIEF SITE LAYOUT, REFINERY AND ASU PROCESS

HIT THEM WITH WOW FACTOR UPFRONT

2 Air Products Public

AIR PRODUCTS

Jazan Economic City

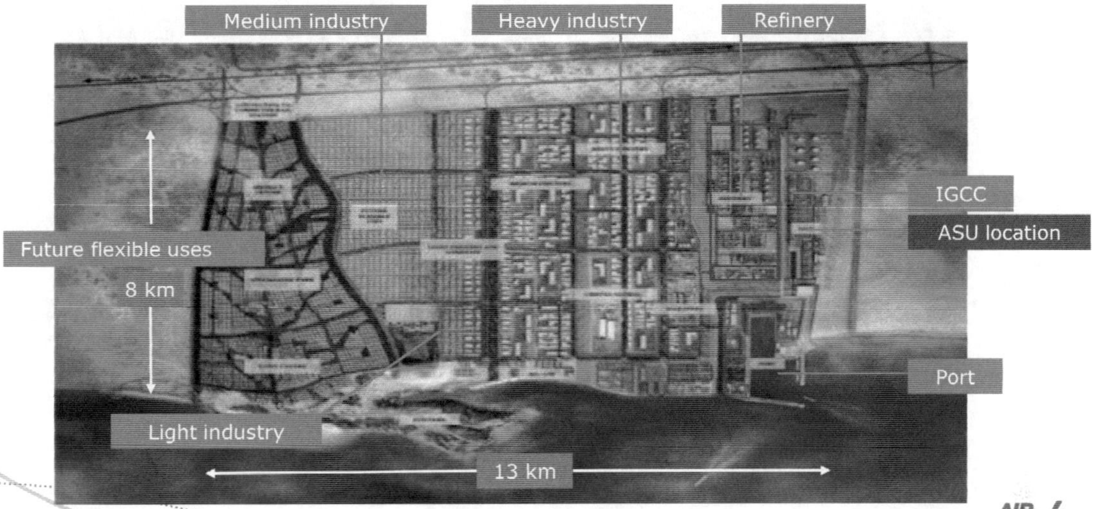

Introduction

- Air Products was awarded a contract in April 2015 to build the World's Largest Industrial Gas Facility to serve the Jazan Economic City located in the South West region of the Kingdom of Saudi Arabia
  - SIX ASU trains producing 75,000 tons per day of Oxygen & Nitrogen to primarily serve the refinery and integrated gasification power plant
  - 20,000 tons of LOX storage and 8,000 tons of LIN storage

LET'S TALK PROCESS SAFETY

Typical Air Separation Process

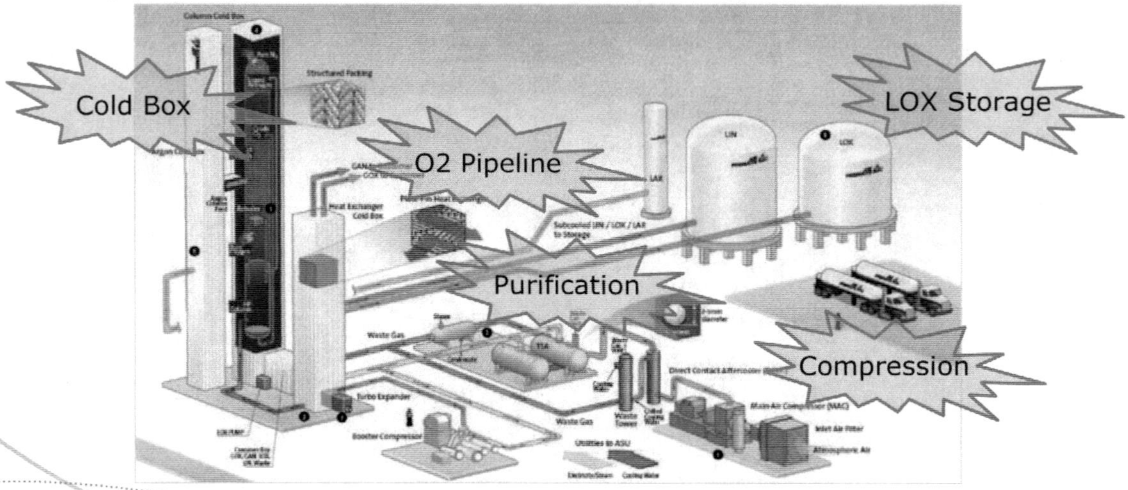

Simplified PFD

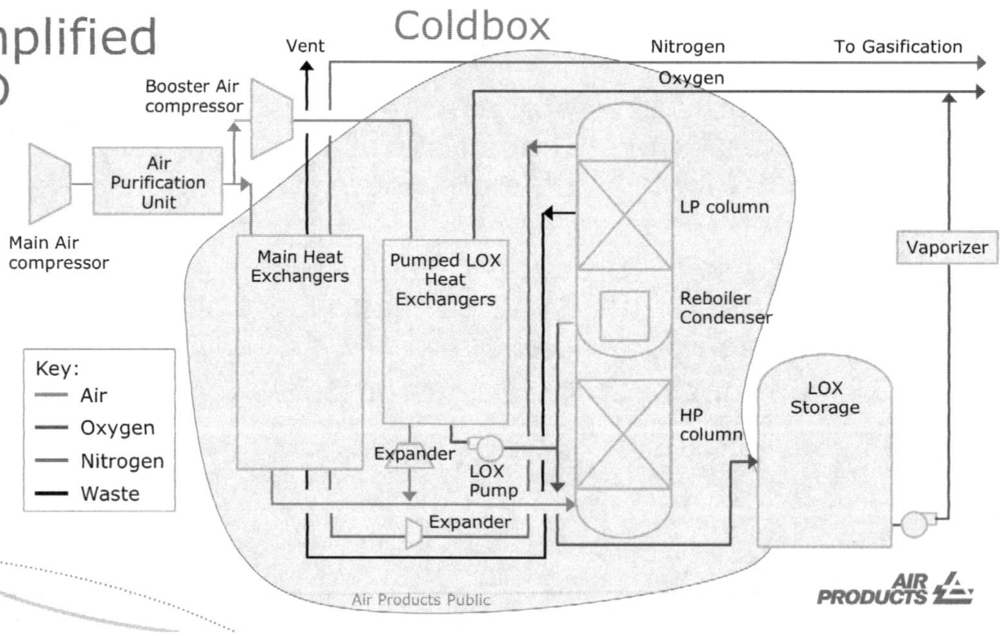

Major Process Safety Hazards & Control for Typical Air Separation Plant Designs

1. Reboiler condenser energy release
2. High Pressure oxygen fire
3. Large scale spill of Liquid Oxygen (LOX)

Where hazards cannot be eliminated by design the intent is:

- To reduce the frequency of the event
- Mitigate for the effects of the event

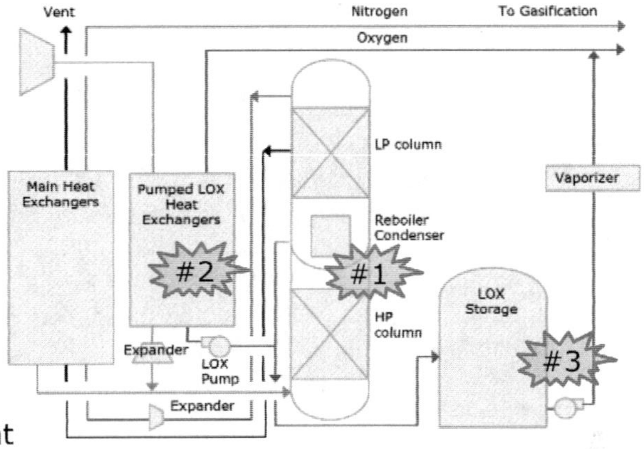

Reboiler Energy Release

- Hazard Mechanism
  - Air contains CO_2, N_2O and trace hydrocarbons
  - Trace amounts of CO_2 & N_2O can enter the process from the air & accumulate & block reboiler passages leading to LOX pool boiling
  - Pool boiling can result in trace hydrocarbon accumulation, which can ignite and initiate an energy release, consuming the aluminum reboiler
- Prevention & mitigation
  - Removal of most air contaminants by front-end purification unit
  - Maintain low contaminant levels in LOX through high purge
  - Monitor CO_2 breakthrough from front-end purification
  - Monitor CO_2 and trace hydrocarbon accumulation in LOX
  - Jazan operates at elevated pressures, increasing trace impurity solubilities and reducing the likelihood of accumulation
  - Jazan maximizes LOX purge via its pumped LOX process selection
  - Jazan will monitor impurity levels in the LOX over time, to identify anomalies early before they can become a problem

High Pressure Oxygen Fire

- Hazard Mechanism
  - Many components in contact with high pressure oxygen have potential for ignition and fire
  - Ignition sources include the presence of oils & greases, impact from particles in flowing gas, mechanical rubbing and heat of compression
- Prevention & Mitigation
  - Material selection of all parts in contact with oxygen
  - Limit maximum oxygen velocities to minimize particle ignition potential. Mesh filters to eliminate particles in sensitive components
  - Oxygen cleaning of all parts in contact with oxygen from source to application
  - Jazan is a pumped LOX cycle which eliminates the known hazard of oxygen compressor fires. A LOX pump cycle is inherently safer than a GOX compression cycle.
  - Vent panels designed into the pumped LOX heat exchanger boxes mitigate energy release due to a high pressure O_2 reaction

Major spill of Liquid Oxygen

- Hazard Mechanism
  - Elevated O_2 levels (> 23.5%) in the air significantly:
    - Increases the chances and severity of a fire
    - Directly impacts exposed personnel in buildings, outside, and in vehicles
    - Impact surrounding process equipment such as air intakes to combustion equipment
  - A spill of LOX will result in rapid boiling and a cloud of cold dense vapor oxygen will hug the ground and disperse slowly
  - Large LOX spills, such as from a LOX tank, can result in hazardous oxygen concentrations several hundred meters from the spill point
- Prevention & Mitigation
  - Lines with high consequence (ie main LOX withdrawal lines from the tank) are located in areas that minimize potential for external damage and have upstream automatic isolation valves to limit release duration should a breach occur
  - Emergency response plans for on-site and off-site personnel established
  - Jazan is designed with bunded areas around storage for containment

Challenges of Multi-Plant Mega Sites

- Common cause failures between multiple ASU plants on a single site.
  - Utilities and overall product control systems may be shared between plants. Loss may lead to total outage, including back-up, and / or loss of ability to maintain safe control.
  - Nitrogen supply is critical, however, it is also known as 'the silent killer'
    - Proper management of Nitrogen use is essential during plant activities
- Planned gas discharge on one ASU plant impacts neighboring ASU.
  - ASUs vent waste gases (low or high oxygen). These may enter air intakes of adjacent ASU equipment / buildings, & worker safety.
  - Simultaneous Operations (SIMOPS) program in place to prevent commissioning/ startup/operations/construction hazards
- Planned gas discharge on one site impacts neighboring site.
  - Combustion processes and seawater demineralization plants vent CO_2, which can enter ASU air intakes.
- Unplanned gas, liquid and energy releases on one site result in over the fence impact.
  - Toxic or flammable vapor clouds from refinery may enter ASU facility.
  - High hydrocarbons from refinery may enter ASU air intakes.
  - High oxygen from ASU may enter combustion equipment in refinery.
  - Fires and explosions may extend within sites or from one site to another.

Oxygen Cleaning on MEGA Sites

- Proper oxygen cleaning techniques for the removal of hydrocarbons & particles for <u>all</u> components in contact with oxygen is critical to avoid potential oxygen fire
- Maintaining cleaning integrity consistently from <u>production end to the end application</u>, as construction management and ownership transitions is a challenge
- Pneumatic pressure test with dry gas preferred, but some sections may be tested hydraulically
  - Drying out lines quickly & effectively is critical to avoid rust & particle generation
- Keep ambient air out of the system after cleaning – use a nitrogen purge
- Specialist training and validation of entire circuit from production point to end user is required

Identification And Mitigation of Hazards

- Reliability analysis focused on managing common utility / control failures. Dynamic modeling of trips (including 'sympathy trips') and back-up response times.
  - Nitrogen used for safety purposes is backed up by systems not requiring power, Cooling Water or Instrument Air.
  - Vaporised cryogenic liquid back-up systems are designed for 'quick response' to ensure the systems are ready for use in emergency situations.
- Hazard Identification used traditional methods such as Hazard Review and Plot Plan Review, but with additional 'nodes' for Facility Siting.
  - Occupied Building Siting Analysis ensured that buildings were designed to withstand the hazards originating from the ASUs and from the neighbouring site.

13 Air Products Public AIR PRODUCTS

Identification And Mitigation of Hazards

- Gas dispersion analysis used to assess the potential for vented gases to enter adjacent plants or impact operating areas.
- Hazard Review augmented by Quantified Risk Analysis (QRA) techniques and modelling to identify on site and off site hazards.
  - QRA identified areas where acceptable Fatal Accident Rates (FAR) might be exceeded on site and further safeguards were added as required.
  - QRA information communicated with neighbours and Individual Risk (IR) reviewed to ensure that risks were acceptable or 'as low as reasonably practicable' (ALARP)
- Fixed and personal gas detectors, blast proof buildings and toxic refuges used to protect personnel.
- Communication of risks and mitigation approaches between neighbors, sharing critical alarms and emergency response "hot lines" between control rooms, along with joint emergency plans and drills.

14 Air Products Public AIR PRODUCTS

Conclusions

- Jazan will be the World's Largest Industrial Gas Facility. The scale and proximity to adjacent operations has required careful Process Safety evaluation
  - Process Safety risks inherent in the ASU design
  - Potential events that impact adjacent operations
  - Potential events in adjacent operations that impact the ASUs.
- Coordinated planning and ongoing communication between neighboring sites will be essential to successful startup and safe long term operation

Thank you
tell me more

Formulating Effective Assets Mechanical Integrity Management Program

- Conference: CCPS Middle East Conference on Process Safety
- Year: 2017
- Time:
  Monday, October 9, 2017 - 4:00pm-4:30pm

Authors:
Aldeeb, A., Siemens Energy, Inc.

Mechanical Integrity experts can articulate the many benefits of being able to effectively manage the process industry assets mechanical Integrity. However, the significance of assets mechanical integrity means many things to different stakeholders; management, operations, maintenance, projects, and HSE, but we all can agree on some common benefits that must be achieved. These include the benefits of maintaining safe operation and contained hazardous components at all times, increased profitability, operation sustainability, earning the license to operate, and grow the business. Did we get here by luck or are we really good and doing things right? Know the why you got here, how easily things can be different, and understand how to sustain a good safety culture.
The ultimate goal of any successful assets mechanical integrity management program is to safely extend the assets lifecycle and maximize the operation efficiency and productivity.

Although many operators spend significant effort to establish such successful programs; however, most recent regulatory and expert audits reveal common mistakes and failures.

In this presentation, developing mechanical integrity management program objectives, strategy, systems and procedures that define requirements, scope, roles and responsibilities, competency, technology enablers, performance indicators, audits and validations and continuous improvements will be presented with specific focus on:

- Strategies and procedures

- Performance indicators

- Sustainable implementation and continuous improvement

- Digitalized technology enablers

A New Approach for Managing Process Risks and the Effective Communication of Those in a 3D Virtual Environment

- Conference: CCPS Middle East Conference on Process Safety
- Year: 2017
- Time:
  Tuesday, October 10, 2017 - 1:15pm-2:00pm

Consequences and risks from unwanted process releases have been quantified for many years using a variety of tools from quick and simple 2D analytical and semi-empirical techniques to advanced 3D simulation methods based on Computational Fluid Dynamics (CFD). A 2D analytical modelling approach does have benefits including calculation speed and ease of use but it is accepted as not being accurate in and very limited when assessing the effects from complex physics such as explosion, fire or dense gas dispersion. The limitations of the semi-empirical tools with regards to the consequence predictions have not however prevented this simplified approach from being widely used to conduct Quantitative Risk Analysis (QRA) studies. One of the drawbacks is that more and more detailed inputs are required to counteract the empirical approximations used in the calculations. These QRA studies have become very complicated, time consuming and difficult to understand. Besides the specific outputs are often difficult to communicate out to non-safety experts.
With a 3D CFD modelling approach, an analysis requires more computer time to run typically but less specific and constraining simplifying inputs by the analyst as the physics are actually modelled instead of being approximated. It makes use of the 3D geometry data, which is also key to producing meaningful results. Any 3D analysis will produce far more comprehensive and accurate results, useful in removing uncertainty. Getting accurate results is crucial to the decision making process when it comes to safety matters, design issues or modification of existing facilities or process. However another additional benefit, often overlooked, is the powerful visualization capabilities from 3D CFD tool such as FLACS®. The realistic outputs makes it easier for engineers to understand the impacts of different options and control measures, which can be quantified and justified. Engineers can reliably evaluate the performance of control measures using the 3D computer model without the time, expense, and disruption required to make actual changes onsite.

It has been very common to post-process CFD results to generate risk based predictions thereby avoiding unrealistic and impractical design solutions based on worst case scenarios. This practice is standard for projects in the North Sea (applying Norsok Z-13) using FLACS® as the tool of choice for CFD consequence prediction. Many companies have developed their own probabilistic approach using FLACS® and implemented it through the scripting facility available in the standard FLACS® software.

There are many excellent applications of why 3D CFD modelling has been be selected ahead of a 2D semi-empirical approach such as: smoke and gas dispersion prediction, validation of hypothesis on accident investigation, location of gas detection, optimization of separation between equipment and also facilities, specification of passive fire protection, location of safety equipment or HSECEs, etc.

Gexcon, the developer of FLACS®, recognised the need for a standard risk tool based on and integrated with the FLACS® pre-processor, solver and postprocessor to optimize the setup, simulation runs and interpretations of the results. As a result the FLACS-Risk add-on has been developed to address the needs of engineers who are seeking to calculate accurately the baseline risk picture for a given design option and quantify the impact of mitigations. FLACS-Risk can not only produce exceedance curves at key locations based on multiple scenarios but it allows users to get much more insight in the anticipated risk through extensive 2D and 3D risk visualization. Being able to setup multiple runs quickly, solve in parallel, post-process the results efficiently and compare the impact of design options in a very graphical and intuitive interface is key to speed up the evaluation of risk reduction measures and thus improve the design of assets.

FLACS-Risk benefits from the powerful post-processing capabilities of FLACS permitting detailed visualization of multiple results in three dimensions.

Predicting accurately and improving the understanding of the process risks are key Gexcon objectives with FLACS-Risk but Gexcon has also developed an environment where those risks can be communicated effectively to a much wider audience, following any analysis. Gexcon has developed an interactive 3D Virtual Environment (VR) where the FLACS® analysis results can be coupled and displayed in a surrounding matching the real physical plant facilities. This 3D hosting environment can be used for a wide range of applications from reviewing risk analysis work as part of a project to training staff on site e.g. emergency preparedness and site evacuation.

Gexcon will present both the latest developments of the FLACS-Risk tool for risk predictions and show an example of an integration of FLACS results into a VR application for an onshore facility.

A Streamlined Approach for Full Compliance with SIL Implementation Standards, which has been implemented successfully around the world

- Conference: CCPS Middle East Conference on Process Safety
- Year: 2017
- Time:
  Tuesday, October 10, 2017 - 1:15pm-2:00pm

Authors:
Bridges, W., Process Improvement Institute, Inc.
Many companies put FAR too much redundant effort into determining what SIL is needed and then verifying the SIF design will give the SIL targeted. This paper shows how to apply the qualitative definition of IPLs within the setting of a process hazard analysis (PHA) to get most of the gain from *LOPA* without doing a LOPA (without using numerical values). We show the way we use a PHA team to identify when a SIF is needed and to select the proper target SIL. This portion of the SIL evaluation and the identification and labeling of the IPLs during the PHA/HAZOP does not take any longer than a normal PHA/HAZOP, once the right habits are established. Note that this approach eliminates the need for a separate SIL Evaluation Study to identify the SIFs and select the target SIL. Then, this paper describes how to perform the SIL Verification and Safety Requirements Specification (SRS) remotely, again without the need for a redundant team meeting. This approach has been used at many sites and for thousands of SIFs.

Change Impact on Relief System Design (Understanding equipment impact and the evaluation warranted)

- Conference: CCPS Middle East Conference on Process Safety
- Year: 2017
- Time:
  Tuesday, October 10, 2017 - 1:15pm-2:00pm

Various scenarios can cause the pressure within equipment to exceed design pressure. Appropriately sized relief devices are one of the most common methods of protecting equipment and personnel operating this equipment. A key challenge to being in compliance is the dynamic nature of processing plants. Numerous changes can be expected to happen over the lifecycle of a processing plant. Management of Change, one of the Process Safety Management (PSM) elements, requires equipment impact evaluation when these changes occur to ensure adequate pressure relief. This presentation talks about numerous types of changes which will impact equipment and warrant overpressure protection re-evaluation. Additionally, a case study is presented where gaps in MOC process along with other deficiencies resulted in a catastrophic failure of a reboiler.

Changes to Industry Guidance for Relief and Blowdown System Design and Impact on Existing Infrastructure

- Conference: CCPS Middle East Conference on Process Safety
- Year: 2017
- Time:
  Tuesday, October 10, 2017 - 1:15pm-2:00pm

Authors:
Frey, P., PSE limited

As the last line of defense for process and equipment integrity, relief, blowdown and flare systems need to be adequately designed and maintained. Systems need to be periodically re-assessed so that they operate safely throughout the life of the facility that they help protect. This talk highlights the importance of ensuring flare and relief systems remain fit for purpose throughout their lifetime maintaining accountability for plant modifications, changes to operations, and changes to industry guidance and regulation. Recognizing high profile an recent incidents involving depressurization, relief and flare systems, including Westlake, Grangemouth and Piper Alpha and others which that have led to loss of containment due to excessive vibrations and brittle fracture of the flare piping, we show that the root cause can be linked to failure to recognize hazards and perform adequate analysis, which is fundamental to Process Hazards Analysis programs and Overpressure systems verifications.

In recognition of these & other incidents, we describe how relief and blowdown adequacy assessment has tightened in recent years in regards to significant changes to API 521, and in particular:

1. the importance of verification of relief and flare systems to ensure they comply with latest Industry Standards

2. to ensuring the suitability of materials of construction so as to avoid brittle fracture risks during depressurization

3. the adoption of the more rigorous analytical methodology for assessing vessel survivability under fire attack for blowdown system design

4. to assessing the likelihood of failure due to acoustic and flow induced vibration in flare piping

Through a number of case studies we describe a methodology for assessing and analyzing existing infrastructure for these risks and explain how a detailed model-based analysis can ensure an inherently safe relief and blowdown system design that is compliant with the latest industry guidance.

Continuous improvement toward excellence in PHA studies

A PHA/risk assessment methodology focused on **high risk** processes utilizing SIS to **reduce** process safety risks

1.
Evolving from prescriptive requirements to **Risk Based** approach.

2.
Sharing a practical approach, under the **Risk Based concept** applying Process Hazard Analysis (PHA)/risk assessment methodologies.

3.
When to go for the **extra mile** in terms of Risk Assessments to demonstrate **due diligence,** achieving **cost effective** designs.

Risk Based methodology for SIS (authors' view):

A SIS design following the risk based approach shall demonstrate, with the appropriate confidence level, due diligence in the application of **Process Safety Risk Management**.

Fundamental risk assessment tools are applied to define the Safety Integrity Level (SIL) of the Safety Instrumented Function (SIF) as an Independent Protection Layer (IPL), determining the level of scrutiny needed for further application of more complex risk assessments analysis.

This will lead to a SIS appropriately designed to meet the specific requirements of the process/application defining: architectural constrains, testing intervals, common cause failure factors, among other SIS design parameters.

Continuous improvement toward excellence in PHA studies

Qualitative Risk Assessments:

"Based primarily on description and comparison using historical experience and engineering judgment, with little quantification of the hazards, consequences, likelihood, or level of risk." (CCPS Glossary)

Semi-Quantitative Risk Assessments:

"Risk analysis methodology that includes some degree of quantification of consequence, likelihood, and/or risk level." (CCPS Glossary)

Quantitative Risk Assessment (QRA):
"The use of Quantitative Risk Analysis results to make decisions, either through relative ranking of risk reduction strategies or through comparison with risk targets." (CCPS Glossary)

Prescriptive

Advantages:
-Cookie cutting approach
-Standardized design
-Gold platting solution

Disadvantages
-Costly solutions
-Non-flexible designs
-Over/under design

Risk Based

Advantages:
-Tailored to risk exposure
-Cost-effective
-T&I optimization

Disadvantages
-Advance knowledge
-Risk of lengthily studies
-Challenge the status quo

Risk assessment categories	Risk assessment tools used (typical)
Qualitative	HAZID, HAZOP, What if?, FTA
Semi-Quantitative	LOPA , FTA, RBD
Quantitative	QRA (FTA, ETA, CM, F-N, IRPA)

Continuous improvement toward excellence in PHA studies

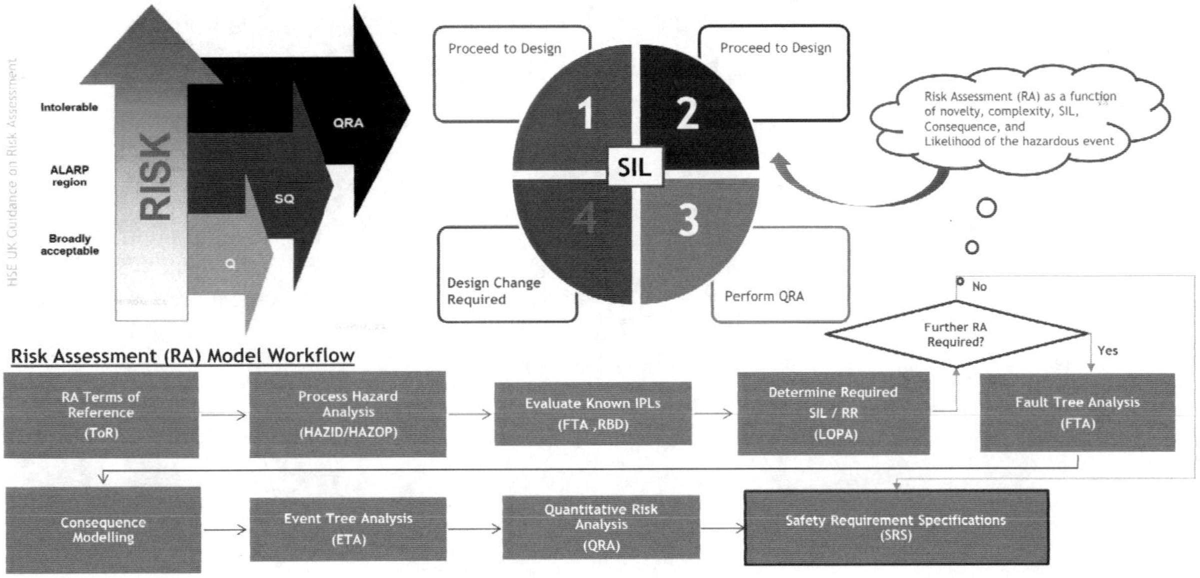

Presenters: Pedro Mujica & Herman Cipriano, P&CSD/PASD/HIPSU

أرامكو السعودية
saudi aramco

Controlling Risk during Major Capital Projects

- Conference: CCPS Middle East Conference on Process Safety
- Year: 2017
- Time:
  Tuesday, October 10, 2017 - 1:15pm-2:00pm

This paper describes the best practices for scheduling and performing PHAs/HAZOPs, etc., during various key phases of major projects. The paper outlines the scope and content of each project phase hazard review and what the outcomes should be. The concepts are not new, but many contractors and vendors around the world do not seem to understand the concepts well. The paper presents the basics, and then shows best practices and examples from various companies. The paper also outlines how information related to process safety should be developed during and then delivered from a major project.

Emergency Planning and Response:
A challenge in process safety management.
Lessons to learn from the oil and gas industry.

Oscar Mauricio Barajas Pinzon

Loss Prevention Engineer

Saudi Arabia

Introduction

- Inadequate or poor emergency planning or response has been a recurring finding in the U.S. Chemical Safety and Hazards Investigation Board (CSBs) in their process safety accident investigations, specially in the oil and gas industry.

- Recently in 14 of the world process safety major incidents since 1998, there were 7 common findings related to emergency response plans and preparedness in fields like training, communication, coordination and logistics. It reflects that emergency management is still being a world wide challenge that requires huge efforts of improvement from managers, insurance companies, internal and external first responders.

Process Safety Incidents with major findings in Emergency Response Plans ERPs:

2013 – West Fertilizer Explosion and Fire.
2011 – Hoeganaes Corporation Fatal Flash Fires.
2009 – CITGO Refinery Hydrofluoric Acid release and fire.
2008 – Bayer CropScience Pesticide Waste Tank Explosion.
2007 – Little Propane General Store.
2006 – Universal Form Clamp Co. Explosion and Fire.
2004 – MFG Chemical Inc. Toxic Gas Release.
2003 – Honeywell Chemical Incidents.
2003 – DPC Enterprises Glendale Chlorine Release.
2002 – Georgia – Pacific Corp. Reactive Chemical Explosion.
2002 – First Chemical Cops. Reactive Chemical Explosion.
1998 – Herrig Brothers Farm Propane Tank Explosion.

ERP Findings and Recommendations:

- The majority of the recommendations relating to emergency response can be categorized in the following topics:

1. Training for emergency responders, including hazardous materials training.
2. Local emergency planning and Community response plans and teams.
3. Use of Community notification systems.
4. Use of incident command system and the Nacional Incident Management System.
5. Conducting emergency response exercises.
6. Information sharing between facilities, emergency responders and the Community.
7. Communication during emergencies.

Emergency Planning and Response

Inadequate or poor emergency planning or response has been a recurring finding in the U.S. Chemical Safety and Hazard Investigation Board's (CSB's) investigations. One of the very early CSB investigations, the Herrig Brothers Farm Propane Tank Explosion, made an emergency response recommendation for training for firefighers, as did the recently closed West Fertilizer Investigation.

To date, 14 CSB investigations have found deficiencies in a community's, facility's or emergency responder's response to an incident at a chemical facility. The majority of CSB's recommendations relating to emergency response can be categorized in the following areas:

- Training for emergency responders, including hazardous materials training;
- Local emergency planning and community response plans and teams;
- Use of community notification systems;
- Use of an incident command system and the National Incident Management System;
- Conducting emergency response exercises;
- Information sharing between facilities, emergency responders and the community; and
- Communication during emergencies (including 911 operator communication, interoperability of radios, etc).

Emergency Response Investigations

Investigations	15
Recommendations	46
Open Recommendations	9
Closed Recommendations	37

In 2009, the CSB released a safety video based on ten years of emergency response recommendations and investigations.

CSB Investigations and Recommendations involving emergency response and planning:

- 2013 - West Fertilizer Explosion and Fire
- 2011 - Hoeganaes Corporation Fatal Flash Fires
- 2009 - CITGO Refinery Hydrofluoric Acid Release and Fire
- 2008 - Bayer CropScience Pesticide Waste Tank Explosion
- 2007 - Little Propane General Store
- 2006 - EQ Hazardous Waste Plant
- 2006 - Universal Form Clamp Co. Explosion and Fire
- 2004 - MFG Chemical Inc. Toxic Gas Release
- 2003 - Honeywell Chemical Incidents
- 2003 - DPC Enterprises Glendale Chlorine Release
- 2002 - Georgia-Pacific Corp. Hydrogen Sulfide Poisoning
- 2002 - DPC Enterprises Festus Chlorine Release
- 2002 - First Chemical Cops. Reactive Chemical Explosion
- 1998 - Herrig Brothers Farm Propane Tank Explosion

Thank You!!

Enhanced Risk Management through Effective Evaluation

- Conference: CCPS Middle East Conference on Process Safety
- Year: 2017
- Time:
  Tuesday, October 10, 2017 - 1:15pm-2:00pm

Unidentified or under estimated process hazards, the way of risk controlling and managing are the major concern in chemical process industries that could lead to chemical release, fire or Explosion. Risk evaluation studies provide organizations with information to help in making decisions for Improving safety and managing the risk of operations. In SAFCO, We usually focus on process Safety issues like acute effects of unplanned chemical releases on plant personnel or the public during all kinds of Hazard evaluations. We typically analyze potential equipment failures, human errors and gaps in the management systems of our organization's process safety program that can lead to incidents. In-line with above SAFCO process safety requirements, we have implement strong program for identifying hazards, analyzing, controlling and managing risks. In SAFCO, we have comprehensive program for management system and the following are risk identification approach and input for risk evaluation:

- Deficiency in process safety information.
- Gaps identified in Operations and Maintenance practices.
- Implement of industrial best practices during gap analysis.
- Training and competency during organizational changes.
- Compliance Audit findings.
- EHSS Violations, Non-conformances & Penalties.
- Process Hazard analysis findings.
- Health Assessment findings.
- Environmental Assessment findings.
- Pre Start-up EHSS Review findings.
- Mechanical Integrity related deferments or findings.
- Management of Change findings.

- Incidents Investigations findings.
- Emergency Planning & Response findings.

All SAFCO employees can initiate risk identification but most of the time it has been initiated by concerned plant operations or respective management system administrator.

The selected, dedicated and competent risk assessment multidisciplinary team has been allotted for the each operating unit and additional member shall be added based on requirements.

The team has to provide appropriate recommendation to reduce risk to tolerable level with respect to corporate risk management procedural guidance during their risk evaluation.

Approval authority, accountability and responsibilities have been set to continue the operating unit with identified risk if the risk mitigation action is not able to implement immediately.

Approval level shall be vary based on level of risk and it is starting from Senior manager plant operations, General manager Operations and Company President.

Periodic audit / field verification is being conducted on all active risk items by the respective risk assessment team and report the gaps to company management accordingly.

Active risk item shall be closed after implementing the recommended mitigation actions and confirmed by respective risk assessment team through field verification.

All active risk items and their evaluations are being maintained by company focal person, which is periodically shared with corporate level, company higher management and stakeholders as part of tracking and closing.

All identified risk and its evaluation with recommendations are being shared with other units/plants risk assessment team of our complex as cross functionalization to evaluate the applicability and corrective actions if required.

This is consider under quality enhancement, knowledge sharing and lesson learnt.

By implementing this approach, our organization benefited as follows

- Enrichment process safety culture,
- Improved training and understanding of the process,
- Reduced consequences of process safety events,
- Improved emergency response,

- More efficient and productive operations,
- Improved involvement of stakeholders,
- Improved process safety documentation,
- Improved operational / Maintenance Excellency,
- Improved regulatory and community relations.

This paper is demonstrate the importance of risk identification, evaluation work practices, procedures, challenges faced during implementation and benefits on process safety.

Improvement in Transient Operation of Ethane Cracker

- Conference: CCPS Middle East Conference on Process Safety
- Year: 2017
- Time:
  Tuesday, October 10, 2017 - 1:15pm-2:00pm

Introduction

A1.1. This report outlines the task of improvement in transient operation of Ethane Cracker for ethylene production at Saudi Petrochemical Co. (SADAF), Jubail Industrial City, Kingdom of Saudi Arabia.

Background

A1.2. SADAF is a joint venture between Saudi Basic Industries Corporation (SABIC) and Shell Chemicals Arabia LLC. It is a petrochemical complex containing six world scale petrochemical plants with a total average output of more than 4 million metric tons per year of various chemicals. The plants include Ethylene, Methyl Tertiary Butyl Ether (MTBE), Ethylene Dichloride (EDC), Styrene-I&II, Crude Industrial Ethanol (CIE) supported by Utilities distribution department (UDD) with production of 12,600 Metric tons per day steam and 250 MW power.

A1.3. Ethylene plant was commissioned in 1986 and had gone through expansion in 1991 and 1997 to an increased production capacity of 1.1 million metric tons per year. The plant receives ethane feed from Saudi Aramco and delivers produced ethylene to other SADAF plants and KEMYA (SABIC affiliate) to manufacture different derivatives. Ethane feed is cracked to ethylene and byproducts in furnaces (crackers). Ethane is fed along with steam to lower the hydrocarbon partial pressure in the radiant coils and to prevent rapid coke formation so as to increase furnace run length. Furnace is followed by primary and secondary transfer line exchangers (PTLE & STLE) where hot furnace effluent is cooled by steam and boiler feed water respectively to stop reaction and avoid unwanted products. The water quench system provides final quench to the furnace effluents before the cooled cracked gas is sent forward to compression and treatment. In addition, a small amount of the heavier components, mainly gasoline, is condensed out in the water quench tower.

A1.4. There are total 12 furnaces, out of which 9 or 10 furnaces are kept in cracking mode. For each furnace, cracking mode is followed by hot steam standby (HSSB) and de-coke modes, to remove coke from inside cracker tubes and subsequently to improve efficiency. This transient operation is repeated for each cracker almost every 180 days depending upon the furnace performance, feedstock availability and production targets. Recently, a fatal incident happened during transient operation of a naphtha cracker. Sadaf management, therefore, decided to carry out assessment and to review the Ethylene furnaces transient operation philosophy with an objective to examine the applicability of that incident at

Sadaf site. A multi-disciplinary team was chartered to complete this critical task, with members from process operations, process engineering, process safety and environment, health & safety (EHS).

A1.5. The team selected the most suited methodology to complete the activity from conception of problem to producing the solution. It included development of activities flow chart, task of data collection, review of past process safety studies, interviews of process operations team, qualitative and semi-quantitative analyses, and proposing possible solutions with actions.

A1.6. First of all, critical incident details were collected. There had been an outside fire from a naphtha cracking furnace which resulted in a fatality. Available information suggests that transient operation (moving from cracking mode to decoke) may have played major role.

A1.7. Teams objective was to analyze cracking furnace transient operation in relation to the above incident. It is to be noted that, owing to number of furnaces, transient operation is almost fortnightly activity which makes it very critical for business continuity.

A1.8. Next step was to collect process information. It included process flow diagrams (PFD), piping & instrumentation diagrams (P&ID), equipment and instrumentation specifications, operation philosophy, standard operating procedures (SOP) startup, shut down, normal and transient, process safety studies like process hazard analysis (PHA) and incidents history.

A1.9. Following critical steps and information were noted during the review of SOPs and data:

a. Furnace effluent line is divided into two lines of different sizes downstream secondary transfer line exchanger (STLE). Larger dia pipe is to line up furnace effluent to quench tower through a motor operated valve (MOV) during cracking and HSSB modes. While smaller dia line is used for aligning the flow to decoke pot and cyclone.

b. Larger MOV gate is of double wedge design. Sealing steam at a pressure higher than that of quench tower side, is supplied in cavity between the two wedges to avoid any valve passing.

c. SOP asks for Double block & bleed to cut off ethane feed to furnace prior to switch over from cracking mode to HSSB mode

d. SOP includes insertion of blind at downstream of 1st block valve of ethane feed line before switching furnace effluent (steam) from quench tower to decoke header (by MOVs).

e. Verification in area and panel is included in SOP for closing larger MOV, opening sealing steam to prevent cracked gas backflow to de-coke header and opening smaller MOV to de-coke header.

f. SOP ensures reading is Zero at hydrocarbon analyzer installed on decoke header, otherwise larger MOV tightness and cavity pressure is checked

g. Each step is signed by panel & field operators on separate copies. After completing the activity, the field operator transfers his signatures on panel copy and wastes field copy (soiled usually).

h. Furnace Feed In / Feed Out is part of job qualification program (JQP) for field and panel operators

A1.10. Important feedback extracted from interviews of field and panel operators included:

a. Field operators, sometimes, face difficulty in closing MOVs. One of either MOVs stick during switchover from HSSB to decoke mode approximately once every ten times. This situation needs to be resolved very fast within 2-3 minutes using hand-jack and motor; or otherwise revert to HSSB.

b. There is a pressure gauge provided to ensure No Backflow during switch over, i.e., closing and opening MOVs in increments one by one.

c. One operator is present in field for normal switchover. If there is problem, more operators are arranged.

d. MOV position indicator to show percent opening is not provided at panel. Only open, close & moving indications are present.

e. All communication between field & panel operators is done via radio which is satisfactory

f. Sealing steam pressure is noted in the field once per shift when applied.

g. There are few opportunities of improvement regarding STLE outlet PG.

A1.11. While reviewing incident history, team came across one relevant incident at site. In addition, root cause analysis and corrective actions were also reviewed for the same incident. Actions found complete and effective.

A1.12. After review of all information, risk assessment was conducted. First part of analysis was qualitative, based on hazard & operability (HazOp) technique following 5x5 risk assessment matrix for the threat of hydrocarbon breakthrough to decoke header,

Risk (R) = Consequence (C) x Likelihood (L)
As released hydrocarbon may catch fire resulting into a fatality (C2) while likelihood was low or very low (L4 / L5).

A1.13. Being fatality scenario, semi-quantitative risk assessment was carried out using layer of protection analysis (LOPA) technique.

$RR = IEF \times PFD_{CB1} \times PFD_{CB2} \times PFD_{CB3} \times PFD_{RM1} \times PFD_{RM2}$ *Where, RR = Residual Risk; IEF = Initiating event frequencyPFD = Probability of failure on demandCB = Control barrier; RM = Recovery measure*
A1.14. LOPA results concluded that cracking furnaces transient operation carries certain residual risk

which needs to be mitigated through various actions.

A1.15. Keeping in view the fortnightly cycle, it was recommended to revise furnace diversion SOP immediately to ensure maintaining MOV cavity pressure once MOV is closed and sealing steam open; adding warning box to sensitize the operator about consequences in case of backflow from cracked gas header to decoke header; and elaborate few steps to keep decoke header side pressure higher than that of quench tower side while opening or closing MOVs. It was also advised to immediately improve JQP resource material by adding detailed discussion of furnace transient operation and consequences of backflow from cracked gas header to decoke header.

A1.16. Team also furnished following short term actions:

a. Complete previous PHA revalidation action related to sealing steam to MOV.

b. Improve hardware and operation regarding STLE outlet PG for a better control as a best practice.

c. As per OSHA standard 29 CFR 1910.119 Appendix A, add hydrocarbon analyzer on de-coke header to safety critical instruments list and define its preventive maintenance accordingly.

A1.17. A long term action was recommended to automate switchover from quench tower to decoke header based on pressure differential transmitter (PDT) across main cracking mode MOV in light of newer designs; as the critical step of furnace switchover is managed manually based on SOP, operator knowledge and field indications only.

A1.18. After taking immediate and short term actions, the company can improve furnace transient operation saving down time as well as reducing a significant risk in a business critical activity to tolerable. Automation requires few long lead items, therefore, can be taken as a separate project, completion of which will save even more downtime and will bring the risk down to as low as reasonably practicable (ALARP).

A1.19. A revalidation / review will be carried out after detailed RCA report of the accident is received from affiliate site.

*By 2017 CCPS MEPSC, improvement results would be available.

Incorporating Risk into the Pressure Relief System Design Process

- Conference: CCPS Middle East Conference on Process Safety
- Year: 2017
- Time:
  Tuesday, October 10, 2017 - 1:15pm-2:00pm

Recent updates to pressure relief system design Recognized and Generally Accepted Good Engineering Practice (RAGAGEP) include provision for risk based overpressure protection. Although risk based approaches are not new, the recent RAGAGEP provides additional guidance and criteria.
Risk based pressure relief and effluent handling system design begins with identification of overpressure scenarios. The risk based approach utilizes existing Process Hazard Analysis techniques (PHAs) and risk tolerability criteria already implemented at Process Safety Management (PSM) covered facilities. When PHAs use Layer of Protection Analysis (LOPA), risk matrix or other qualitative, semi-quantitative, or quantitative approaches, the consequences and likelihood of the scenarios are determined.

The consequence of an overpressure scenario can be predicted using computational techniques, which include consideration for the pressure rise, potential release of toxic and flammable fluids, and resulting vapor cloud due to atmospheric relief or loss of containment. The frequency of occurrence of the overpressure scenario can also be predicted through historical data, or analytical techniques such as fault tree analysis.

Upon determination of the consequences and frequency, the risk can be calculated; which can then be compared to the applicable tolerability criteria. Scenarios with an intolerable risk may then mitigated through a variety of enhancements or safeguards. The cost of prevention, control, and/or mitigation measures are then balanced to ensure a safe and cost effective relief system always ensuring the As Low as Reasonably Practicable (ALARP) principle.

Additional risk factors that are often overlooked during pressure relief system design are also discussed. These include dispersion analysis, thermal radiation, noise, vibration risk, reaction forces and structural supports, metal cold temperatures due to expansion cooling and two phase flow, hot temperatures due to fire exposure and/or runaway reactions, PRV stability, chemical reaction systems, and loss of high pressure/low pressure interface.

Integrating the pressure relief system design into existing PSM practices reduces the effort and cost of developing and maintaining relief system design basis documentation and focuses the designersâ intent on improving plant safety. This risk based approach to pressure relief system design ensures adequate overpressure protection systems and sound investment in human and capital resources focused on risk reduction.

This study presents a practical risk based approach to pressure relief and effluent handling system design and a supporting case study that results in safe and cost effective design.

Inherently Safe Design for Flare Systems

- Conference: CCPS Middle East Conference on Process Safety
- Year: 2017
- Time:
  Tuesday, October 10, 2017 - 1:15pm-2:00pm

Authors:
Irshad, A., Saudi Aramco

Inherently Safe Design is an ideal way to eliminate or reduce Process Safety Risk of a Major Catastrophic Incident. Fully-rated Equipment Design selection is an ideal way to handle many Causes of Overpressure and Embrittlement concerns that may result in a Loss of Containment. However, Designers are advised to use caution and base analysis on a full Process Operating Envelope with respective Safeguard Considerations. The Risks and mitigations should be considered with CAPEX and OPEX before settling on a design. A Case Study will be presented specifically with respect to a Flare Design to show the Understanding of Inherently Safe Design Vis-a-Vis other optimum Designs. A Matrix of Risk, Options and Cost with Each Design will facilitate decision making and selected Design of the Equipment for the Facility.

Process Safety Excellence By Improving the Process Safety Culture within Safco

- Conference: CCPS Middle East Conference on Process Safety
- Year: 2017
- Time:
  Tuesday, October 10, 2017 - 1:15pm-2:00pm

Improving process safety culture is a way to commit to process safety, and the cornerstone to process safety excellence.
Although process safety meaning or concepts exist in SAFCO in 2016 we were able to achieve more than 22 million safe man-hours by applying Risk Based Process Safety Management (PSM) elements mostly integrated with SABIC standards, but implementing PS roles and concept independently enhance Process Safety Barriers and promote Process Safety Culture to identify leading indications.

Process Safety walk-through was a part of awareness to frontline workforce to enhance the sense for what could be leading indications to save the plant and equipment from failure, moreover shearing Process Safety Tips in continues manner to all SAFCO employees help to recognize individual role in Process Safety chain.

Gaps identifying after completing assessment of operating plant parameters, assess safety critical equipment and review and revalidate Standard Operating Procedure, were improved which enhanced Process Safety Management, limit the deviation and improve individualâs behavior and safe plants.

Process Safety Field Engagement (P-SaFE)

Asset integrity, maintenance and inspection become increasingly important concerning Process Safety as plant ages. Process safety requires constant attention. Absence of incidents is not a good indicator that process safety is being managed. Aiming to manage Process Safety incidents by controlling of Process Safety leading indicators. This is a crucial role and process in term of process safety excellence where P-SaFE is part of it.

P-SaFE is a periodic formal verification rounds; mainly done by leadership (Go and See/Walk the Talk); of the health of existing Process Safety barriers using predefined checklists to identify the potential for the design barriers missing and/or defect, which may result in a significant Process Safety Incidents. P-SaFE is a crucial sub-process of ESP Technical pro-active monitoring.

A review of a collection of 130 existing Bow Ties had identified 25 major barriers give 80% of the safeguards. The P-SaFE checklists were made around these safeguards.

P-SaFE at SADAF:

Ø P-SaFE program started at SADAF in 2011.

Ø 5 checklists developed and were refined to suit site specific conditions.

- Equipment & Process Condition
- Relief system
- ESD & MOC
- Operations
- Recovery Measures & Procedures.

Ø P-SaFE program rolled out site wide in 2012.

Ø A management system had been initiated and approved in 2014.

Ø As part of Process Safety continuous improvement, a newly checklists had been introduced earlier and those are about Tanks Operational Safety, Draining and Venting, Fired Heater and Rotating Equipment.

Purpose:

Real Process Safety is out in the plant and not behind our desks. Promote active observation and reporting of field based process safety issues

Demonstrate Visible Leadership Commitment by focusing on major process safety hazards and staffs behavior around that (rather than auditing conditions)

Good interaction via conversations with frontline staff and observations in field. Listen to the concerns of those who are closest to the process

Employees motivation by looking at actions of business leaders.

Resource to assist leaders to have quality conversations asset integrity and PS

Resolving chronic unease and promote process safety culture Increase understanding on Process Safety management via barriers thinking (A way of Process Safety culture promotion).

Training aid for staff to look for abnormalities in the field.

Objective:

The objective of P-SaFE is to prevent incidents resulting from unintentional release of energy or hazardous substances. It is about keeping the materials and product inside the designated containments such as the pipes and vessels. *This is accomplished by Process Safety barriers.*

Expected outcomes:

Ø Effective Process Safety Barriers.

Ø Zero Loss Of Primary Containment (LOPC)

Ø Effective Process Safety performance.

Ø Zero Process Safety and EHSS Incident.

RADIATION RISK & SAFETY PROGRESS IMPROVEMENT

ABDULLAH AL ZAHRANI
Radiation Safety Officer – Saudi ARAMCO

Thursday, November 09, 2017

SEIFALLAH BELHADJ
Quality Manager – Advanced Inspection services

INTRODUCTION

- Radiation sources are used in several sectors such as medical, industrials, and at laboratories.
- Nowadays, the world is moving to build more and more nuclear power plants for generating electricity and the risk of radiation accident is also increasing. In the gulf countries ,the general public is not fully aware about the risks associated with radiation, related effects and types of protections Moreover, there are potential risk and threat associated with using of radioactive materials.

- Establish high security system to protect the radioactive sources from handling by unauthorized personnel
- Establish cooperative radiation emergency response plan between GCC countries.
- Initiate new society related to radiation protection under GCC
- Develop New National Experts in Radiation Protection field

OBJECTIVES

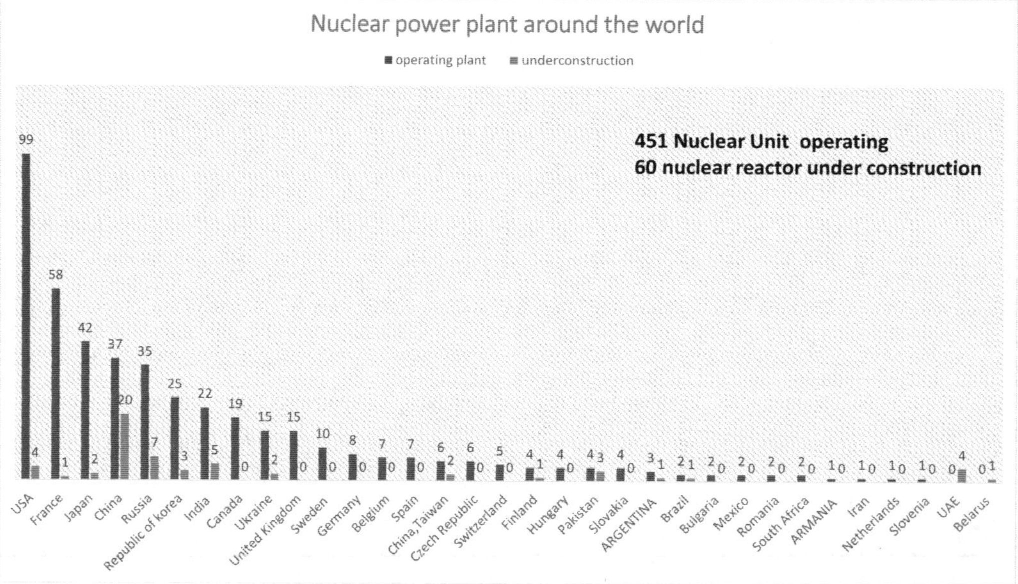

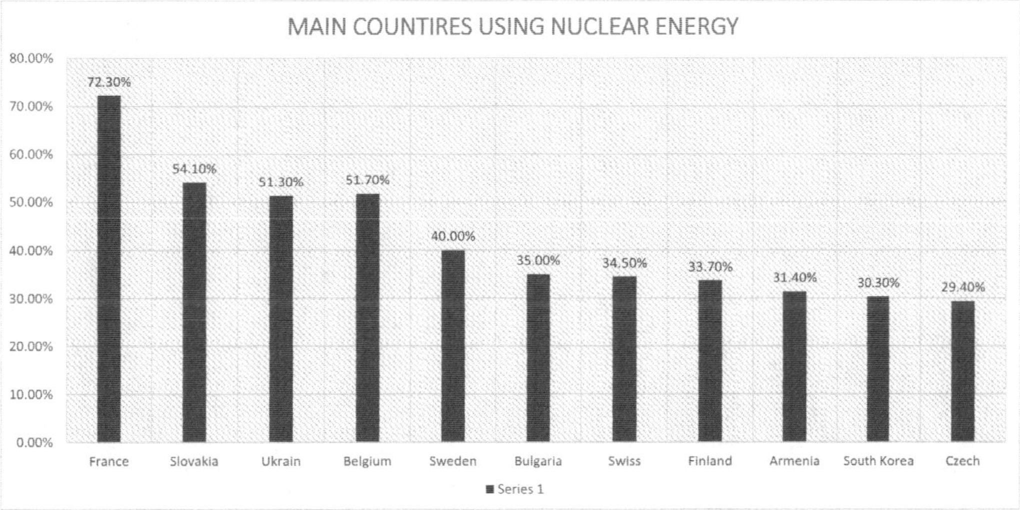

ANALYSIS

Sources of Radiation is all over the world, only if you count the Number of sources available in KSA which is been used for radiographic testing is approximately more than 2000 IR-192 sources only ,other than other sources such as Co-60 ,Cs-137 (approximate calculation),if each of them have an average activity of 20 Curies ,then the total activity available in the KSA will be **40000** Curies, which is a huge amount,

Just Imagine what will happen if such materials are mishandled ,the consequences will be very devastating.

CONCLUSION

Radioactive sources are using with multiple subjects to serve the society and the associated risk can be controlled by establishing effective defenses system against radiation hazard and protect the societies and environment.

Nuclear plants are essential to minimize the power crisis for future since we cannot rely completely on fossil fuels as they may be not available after few years.

Hence we can utilize the advantages of nuclear plants at the same time we have to protect ourselves from the harmful effects of radiation by getting knowledge about radiation and preparing emergency plan in case of any radiation incident, creating preventive measures by using new technologies in construction & design of the nuclear plants.

Representing Human Factors in Bowties as per the new CCPS/EI Book

- Conference: CCPS Middle East Conference on Process Safety
- Year: 2017
- Time:
  Tuesday, October 10, 2017 - 1:15pm-2:00pm

The Chemical Center for Process Safety (CCPS) issues books that have become the *de facto* best practise for process safety management globally. For example the CCPS issued the original LOPA book in 2001 and the Risk Based Process Safety book in 2007. We, the authors and members of the sub-committee advising on the content of the book, are now getting near to the end of a long process to write, develop, review, edit and issue a CCPS/EI Concept book (prepared together with the Energy Institute) on the use of BowTies in risk management. The final chapter of the draft book addresses representing human factors in bowties, part of which is summarised in this paper.
In bowties barriers are in place to stop the threat leading to the âtop eventâ or to prevent the top event leading to catastrophes. These barriers are either passive barriers (e.g. crash barriers, bunds), continuous barriers (e.g. ventilation) or active barriers. Active barriers need to have separate elements to âdetectâ the threat, âdecideâ what to do about it and to âactâ to stop the trajectory. This means that typical human factors aspects such as âtrainingâ, âcompetenceâ, âsafety cultureâ or âleadershipâ cannot appear by themselves as barriers on the main threat line. Instead they are essential to ensure the correct operation of the barriers. As such they appear as safeguards on the degradation factors that affect a barrier.

Almost everyone agrees that a positive safety culture, process safety leadership, etc. are needed for the safe operation of a facility. There is, however, currently no clear way of demonstrating how these aspects prevent the fatalities from a major accident. The approach described in this paper uses multiple layers of bowties to demonstrate the clear link between these human factors and preventing major accidents. Sharing these bowties with operators would support positive safety culture efforts.

There are four important premises addressing human factors in bow tie analysis:

- In basic (âLevel 0â) bowties human error should preferably not be directly modelled as a threat that can lead to a top event. The threats are inherent threats arising from the operation.
- Active barriers, with detect, decide and act componentsÂ very often involve humans in one of the elements and so it is important to understand how these degrade the barrier
- Mechanical failures and human errors can defeat or degrade these barriers. Safeguards are then needed to be put in place to mitigate the risk and to prevent this happening. The term âsafeguardâ is

used to denote those controls not meeting the full requirements of an active barrier.

- Most of the controls that organisations use to minimise human error cannot meet the criteria necessary to be considered active barriers in their own right (effective, independent and auditable). They are, nevertheless, very important safeguards to manage the risk from human error leading to degrading the performance of active barriers. Safeguards can range from local warnings and signs, the design and implementation of alarms and the human machine-interface to control systems, through Job Design, operating procedures and cross-checking practices, to Stop Culture (the willingness of front-line personnel to stop work if they have any concerns over safety), Leadership, etc.
- More detailed bowties (referred to in this paper as âlevel -1â, âlevel -2â, etc.) can be developed focusing on understanding how those safeguards, the human and organizational controls, can themselves be degraded or defeated. Typically two levels will be needed (i.e. the top level, which identifies the main barriers, and level -1, analysing human error as a degradation factor for the top level barriers).

The paper uses a Buncefield-type scenario of overfilling a fuel storage tank as an example to illustrate the application of this approach.

Risk Assessment for Increase in Plant Capacity

- Conference: CCPS Middle East Conference on Process Safety
- Year: 2017
- Time:
  Tuesday, October 10, 2017 - 1:15pm-2:00pm

In a competitive working environment of chemical process industry, it is often required to maximize the production capacity based on the available plant capacity margins or some time going beyond the design capacity by taking into account the cushion available to increase the load as per process conditions without resorting any specific debottlenecking studies. In this exercise the financial gain is a obvious factor but the sustain operation at the elevated plant load (especially beyond design capacity) warrants a structured Risk Assessment for the whole process plant / equipment / machines to assess the hazard and risk to avoid any process safety related incident.
Different Risk Assessment approaches can be used in such scenario like HAZOP, What IF Analysis etc., but typical HAZOP guide words / deviations cannot provide the complete coverage of risk related issues as this is a very special case where all aspects of risk associated right from human factor to catalyst performance needs to take into account. For such cases, one of the approaches adopted is taking into account the layers of protection approach. This structured manner for each selected node with its different layers of protection, will help to check and assess any gaps to bring the overall process at acceptable risk level, while being operated at higher capacity load.

Following layers of protections are taken into account for Risk Assessment study:

BPCS (Basic Process Control System)

- Review of existing Alarm levels
- Control valves (output and response)
- Load margins with pumps/motors/prime movers
- Change in set points for controllers, alarm switch etc.
- Chemical Kinetics at elevated load (runaway reactions)

Equipment (pressure vessels, columns, HEX, pipelines etc.)

- Review of design data sheet.
- Past records of reliability and Turnaround history for equipment
- Erosion issues due to high throughput
- Review of Pressure drop across HEX, columns etc.

ESD / SIS

- Review of existing ESD/SIS data and requirement to change set points due to elevated plant load conditions.

Mechanical Protection system

- PSVs / RD capacity as per the design data sheet and elevated load conditions

Human Factors

- Operators response time in emergency
- Change in operating procedures and trainings
- Upgrade of Emergency Response plan

For a selected node or for any critical piece of equipment like Reactor, all above mentioned factors to be assessed and compared with the design parameters / simulation study and conclude that if the Risk is under acceptable level or not. All relevant process streams to be checked and assessed as per the above mentioned criteria to ensure all hazards and associated risks are taken into account while the plant operated at elevated load capacity.

This approach has been very useful ensuring that hazards and risks associated are taken care with increase in plant load, especially beyond design capacity. This approach can be used with any risk assessment model like What IF analysis or HAZOP while checking each and every parameters (as mentioned above) with available safeguards or recommendations.

ISTITUTO ITALIANO DELLA SALDATURA
Il Gruppo

The Importance of Feedback from Failure & Root Causes Analysis in Risk Based Inspection Studies

M. De Marco – Istituto Italiano della Saldatura Group

1. The Dynamic Process of Plant Risk Based Inspection - RBI

- The RBI approach to reduce risk by optimized inspection strategies is a very dynamic process in the complex world of the Asset Integrity Management of process plants.
- RBI allows the use of models to define Likelihood of Failure (LoF) and Consequence of Failure (CoF).
- Failure analysis of components after premature failure event can be the basis for RBI real-time replanning.
- RBI → Expert tool for expert people

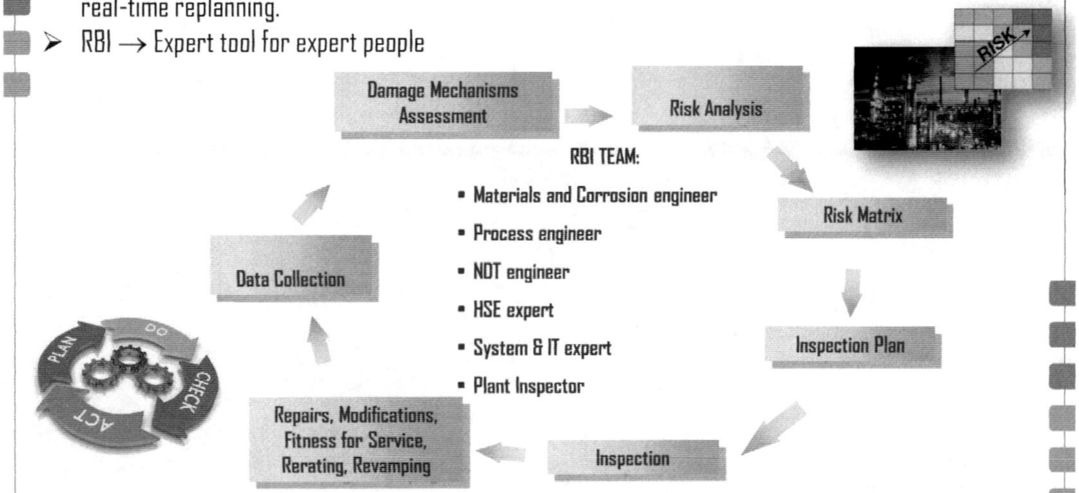

ISTITUTO ITALIANO DELLA SALDATURA
Il Gruppo

2

2. Failure Analysis

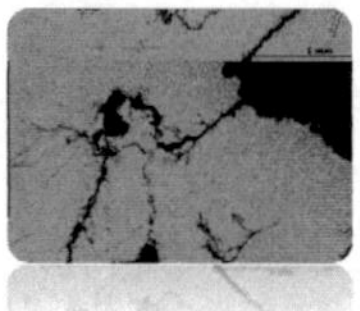

- Damage event (e.g. corrosion driven)
- On-site inspection
- Samples cut-out
- Interview and data collection
- Laboratory investigation
- ***Identification of damage mechanisms → Feedback to RBI team***

Physical causes and mechanisms

3. Root Cause Analysis

- Causal factors schematization (clear & latent)
- Barriers and organizational failure
- "Root causes" identification
- ***Lesson learned → Feedback to RBI team and plant management***

Latent causes & and organizational failures

4. RBI, FA and RCA as Process Plant Risk Management Tools
BowTie Diagram

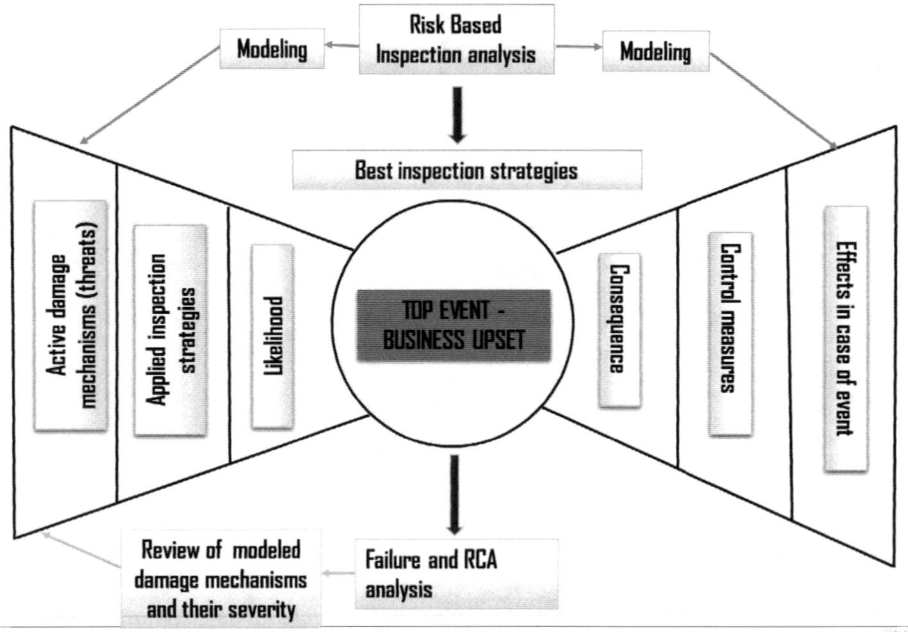

The Reality of Risk | AIChE Academy

- Conference: CCPS Middle East Conference on Process Safety
- Year: 2017
- Time:
  Tuesday, October 10, 2017 - 1:15pm-2:00pm

Authors:
Jones, S., Petrotechnics Ltd

Intro

At the frontline of any hazardous industry, production happens, activities occur, and operational risk and human interaction collide. In the midst of this instability sits hazard identification and risk analysis. Getting it right, making it operational, is critical.

Organisations are tirelessly working to redefine their approach to risk management. The real challenge is arming everyone in the organisation with the right information at the right time so everyone can make the most effective operational decisions. McKinsey and Co. reports, profound technological advancements are disrupting old ways of working in oil & gas and are enabling step changes in safety and productivity. In fact, Accenture reports that companies that understand digital transformation earn 26% more profit than others. The hazard identification and risk analysis of tomorrow will experience massive disruption from digital innovation.

To enable better decision making, people need to be better informed and connected. Whatâs required is a simpler approach that connects disparate processes and people in a way that allows everyone to visualise and manage risk and operations in a new way.

In this paper, we will explore an operational excellence software platform implemented by several major oil companies which has at its core a unique Hazard Identification & Risk Analysis capability. By connecting, translating, aggregating, and visualising disparate systems and data through a common currency we can unlock the potential in data thatâs typically hidden away in silos. We can see and understand the meaningful relationships between previously unrelated data sources. Connecting this information to operational reality will expose the actionable insights that have been lacking in risk management. The result will be reduced risk, improved productivity and lower costs.

Putting Risk in Context

Your operational risk landscape changes as fast as your operational reality. Frontline operations is where process safety management (PSM) happens. Itâs where your people are operating the plant, doing maintenance work, introducing new hazards and potentially impacting the state of the plant. By its very nature, itâs a dynamic and inherently hazardous place to work and manage.

In order to improve risk decision making, we need to understand what is happening, why it is happening, and be able to predict reliably when it will happen. The key is data â connecting everyone with the correct data, in the right context, and using commonly understood language.

Conceptual models like James T. Reasonâs Swiss Cheese and the process safety Bow-Tie methodology have been developed, in an attempt to describe the inter-related nature of operational risk. These work well in a design phase, but the industry has yet to develop operational systems that enable routine management of operational risk on a day-to-day basis for anything other than major, single-source risk scenarios. Attempts to do so fall into the trap of complexity, or result in generalized historical performance indicators - which do not help inform decisions on current or future risks. What is required is an approach that enables hazardous industry operators focus on the elements of risk that are most important.

Any simple risk management system needs to manage major accident hazards, barriers and barrier health.

- Major Accident Hazard (MAH) - The source of a threat of SERIOUS danger, eg. a fire or explosion on an oil and gas facility.

- Barrier - A means of preventing the communication or progress of a MAH.

- Barrier Health - A measure of barrier effectiveness.

A New Leading Indicator of Operational Risk

Although the effectiveness of a barrier (barrier health), is an important indicator, it does not tell the whole story. What is required is a means of measuring how failures in barriers can connect to allow a hazard to become a serious incident. These connections are hazard pathways, a major accident hazard (MAH) pathway, is a potential route a MAH could progress through a series of barriers towards becoming a serious incident.

The ability to monitor MAH pathways is the key that relates the collective performance of a companyâs risk control systems to the real cumulative risk in the operation at any one point in time. This creates a common currency, allowing us to create a holistic view of operational risk irrespective of the source of these risks. This provides the insight executives are looking for by making available a new leading indicator of operational risk that reflects the operational reality of their plants. This approach answers the question of how organisations can gain visibility into the control of operational risk. Incorporating a simple elegant risk model into an operational excellence platform is key to engaging the whole organisation.

It is true that when most major incidents occur, there is a concurrent failure of several barriers. The degree of impairment, and the number of successive barriers compromised, represents a level of risk. But that risk only matters if it is set to occur at a certain time, in a certain area. That is why the approach to operational risk management must be dynamic. Key to the solutionâs effectiveness is the unique visualization of data in real time, which did not exist before. This distils to effectively three different views of the dataâtime, space and risk.

- In space, an operator can see the facility layout with icons indicating how activities and impairments align, changing the risk levels. Highlighted areas represent high risk for current and future shifts.

- In time, traditional Gantt charts display sequenced activities, deviations and scheduling conflicts. Indicators of high or medium levels of risk are shown for current and future shifts.

- In risk, for each shift, the areas where high or medium risk is predicted are shown, along with a picture of the specific barriers impaired. Tools to âsimulate fixâ enable quick testing to see the impact of deciding to repair specific impairments or to defer activities.

All are examples of actionable insights, where data is presented in a way that supports decisions on activity commencement, and schedule priorities and facility operations across the business. And thereâs value for all.

- Frontline operations are provided easy-to-read, data-rich status of equipment and activities, with context to make better-informed, risk-dependent decisions during the shift.

- Site management and planners looking further ahead have access to tools that show the risk implications of scheduling decisions, protracted deviations and planned activities, as well as âwhat-if â scenarios.

- Asset leadership can see, for the first time, risk levels, facility productivity and plan-attainment trends. They can compare asset performance and, where necessary, dive into the data, to see what is generating the highest levels of risk.

- Executives can look across the enterprise and compare asset performance, gaining greater insight into how risk is managed across the whole organization.

A Practical Walkthrough

Work activities are planned and stored in Maintenance Management Systems (ie SAP, MAXIMO, PRIMAVERA). By employing proven integration methods, these activities are imported into the operational excellence platform. Work activities fall into several scenarios; a typical example would be âHot Workâ, which can take place at grade, at height or within a confined space. When defining the hazards associated with an activity, the platform requires analysis of risk associated with the task itself, those related to the location and those relating to any process safety deviations; all building to a cumulative risk. The key here is the risk assessment template â the onus is no longer on the work director to try and remember what hazards are likely to affect this activity. The template has already been prepared and approved and embedded in the software and the selection of location highlights hazards pre-programmed for that location. Finally, a risk assessment is performed (using a configurable risk matrix) to show the impact on the risk level. No progress is possible unless mitigation controls are applied bring the risk level to ALARP. The platform allows previous Risk Assessments/Job Safety Analysis to be reused for improved efficiency but requires a review and acceptance before it is valid for use.

Additionally the provision of Management and Risk Control Systems that have associated KPIâs to show senior managers how well the organisation is performing are available. These are used to measure system health and traditionally illustrate whether elements of the programme achieve or exceed the targeted performance. This allows resources and budget allocation to be changed to impact performance in the next quarter by helping senior leaders focus resources on the areas of the business most exposed.

Conclusion

An operational excellence platform can connect disparate systems and data to harness the potential of people, process and technology to produce actionable insights that can improve risk decision making across the organisation. This is a 21st century approach to hazard identification and risk analysis in industrial operations to reduce risk, improve productivity and lower costs.

Making risk mitigation an integral part of efficient and effective daily operations allows everyone to both visualise and manage risk and activity in virtually real time. With this empowered approach, the dynamic reality can be managed by Space, Time and Risk views to achieve excellence in Hazard Identification & Risk Analysis.

Transforming Towards a "Risk Thinking" Organization By Implementing RBI

- Conference: CCPS Middle East Conference on Process Safety
- Year: 2017
- Time:
  Tuesday, October 10, 2017 - 1:15pm-2:00pm

Authors:
van Driel, C., TÜV SÜD Industrie Service GmbH
Kauer, R., TÜV SÜD Industrie Service GmbH
2017 Middle East Process Safety Conference

9-11 October 2017

Bahrein

Call for PapersTitle:Transforming towards a Risk thinking organization by implementing RBI Authors:

Coen van Driel
Kienbaum Management Consultants
Herikerbergweg 183
1101 CN Amsterdam
Netherlands

Robert Kauer
TÜV SÜD Industrie Service GmbH
Westendstr. 199
80686 Munich
Germany

Abstract:
Case study: RBI implementation at a petrochemical site in Europe

The base for the implementation of a risk based inspection methodology for optimizing risk performance was a process safety assessment at a site in Europe owned by a multinational company operating various petrochemical and chemical plants. The aim of the process safety assessment was to identify the blind spots in the barriers that prevent incidents and to create awareness in the organization of the improvement potential. In the area of asset integrity a lot of good work has been done that is based on the experienced and knowledgeable people in the plant. But two main challenges were identified, risk thinking in all aspects of asset integrity and closing the loop and the link to other disciplines to learn from the output of the process. For example, addressing not only the statutory requirements for inspection but also taking into account the risk of each relevant degradation mechanism and linking in the RBI process the degradation mechanisms with the operating window.

By showing the potential increase in safety and getting a better insight in the asset performance, the sense of urgency for change was created and the will to change established.

Before going into the details of realizing RBI transformation what is RBI? Risk Based Inspection (RBI) is an integral part of a holistic asset integrity management. The assessment determines the risk of a loss of containment for static equipment, by determining the probability of a failure induced by identified relevant degradation mechanisms and the safety and business consequences of such a failure. The aim of the RBI program is to achieve an optimal inspection program on the basis of the potential risks.

Sponsored by the Site Leadership team the RBI implementation team started with the transformation project. The goal is to upgrade the inspection program with a risk based approach to accomplish an even better integrity of the static equipment on site and to have tailored inspection plans the address the relevant degradation mechanisms by bringing in the knowledge of a multidisciplinary team of mechanical, inspection, process, corrosion and operation engineers.

The target of this project was to install an RBI program and at the same time to overcome typical pitfalls during RBI implementations:

· RBI is only an inspection activity and not linked to knowledge about degradation mechanisms/actual process conditions

· no implementation/ RBI master plan -> aimless or unrealistic implementation time/everything at the same time

· Sense of urgency: many RBI projects are done as a tick in the box and not with a thorough understanding of the value for the organization. The results, RBI recommendations, are not fully realized in the organization

Unique aspects of the implementation: a combination of human, process and technical transformation expertise are part of the project to make a sustainable implementation possible.

The success of the RBI Development phase is based on an alignment between reliability engineers from the project, experts from the companies headquarter as well as TÜV SÜD / Kienbaum to establish a common understanding of RBI processes and tools among the participants from different regions. Although the essence and the goal of RBI is the same the interpretation of certain aspects of RBI can differ across the regions as well as the regulatory boundary conditions.

The approach is based on the in-depth technical and RBI knowledge and human/system driven transformation knowledge to make it happen.

AS RBI is one of the key columns of asset integrity and being the backbone for Process Safety it is an excellent way to introduce Risk thinking in the organization. The behavior is in the organization is gradually changing towards a risk thinking behavior based on the implementation of RBI followed by transforming RCM, SIL, do maintenance etc into risk based way of working. As the processes of Process safety (see also figure 1) are interlinked as the beams of a raft the risk based way of working is spread to the processes as operating window, safe work processes, etc. In the end bringing risk thinking into the core of the organization to enable an better performance in reliability and process safety.

Figure 1: The Risk at the Core of the Organization model (Kienbaum TuvSUD) enables a focus on all aspects which results into a high performance.

Structure critical risk, incident and audit data on bowtie barriers

CGE
Risk Management Solutions
Making risk understandable

Paul Haydock
Product manager
p.haydock@cgerisk.com
www.cgerisk.com

In short…

The problem:

Engineers and managers can suffer from information paucity, overload, delayed delivery or indigestible and unrelated data.

The solution:

Combine risk, audit, incident and equipment state data onto bowtie barriers for a fourfold view of barrier management.

BowTieXP &
IncidentXP
AuditXP
CGE Risk Management Solutions BV www.cgerisk.com

There's Beauty in big data

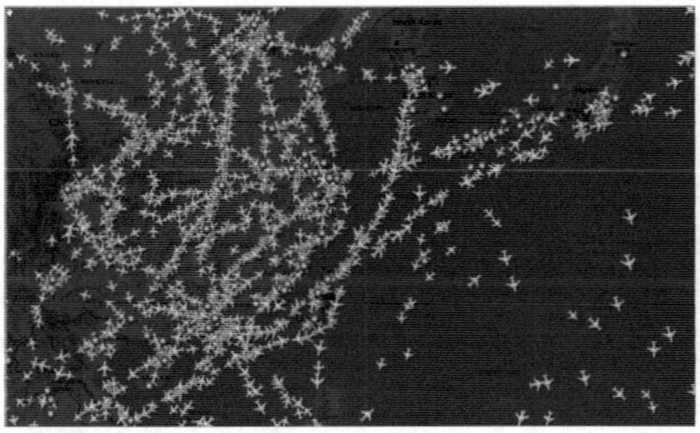

CGE
Risk Management Solutions

3

When data gets ugly

DATA:
Incidents
Audits
Equipment
MoC
PTW

Cause

- Data paucity or overload
- Not verified / verifiable
- Poorly structured
- Incongruous sources
- Indigestible
- Meaningless / misapplied

Effect

- Poor decisions
- Repeated incidents
- Production delays

CGE
Risk Management Solutions

4

Big Data

Failure

- 1958, Int'l Geo Year project. 1000s of researchers

Photographs | spectrograms | maps, observation logs | tables | charts, documentary films | technical reports.

> too much data, too many platforms

Success

- "1840, fossil classes... Bronn realized... to order chaos = to visualize information and so learn something from it."
- Also, Linnaeus

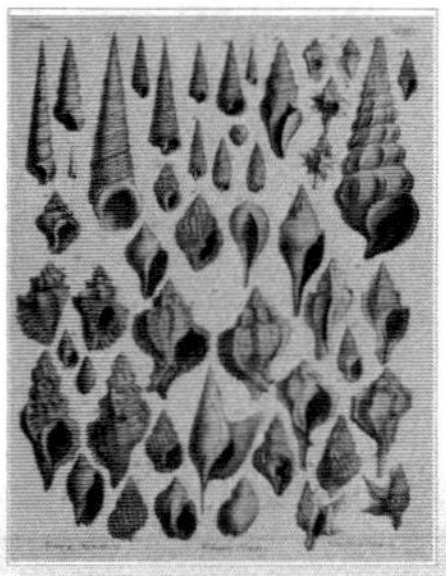

https://www.chemheritage.org/distillations/magazine/information-overload

CGE
Risk Management Solutions

Diverse sources and voices?

Audit data | Risk analysis data | Equipment live state data | Incident

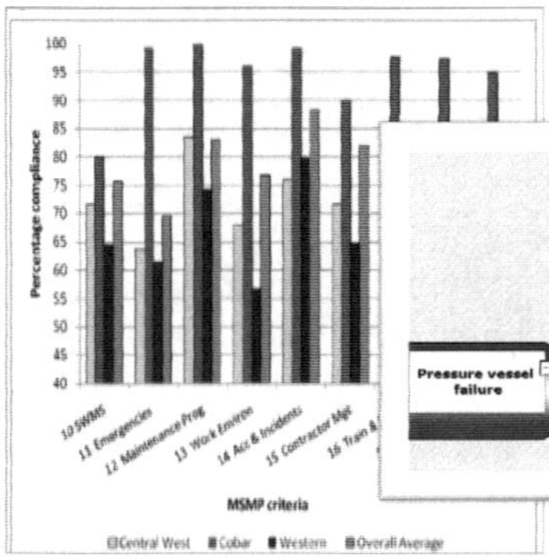

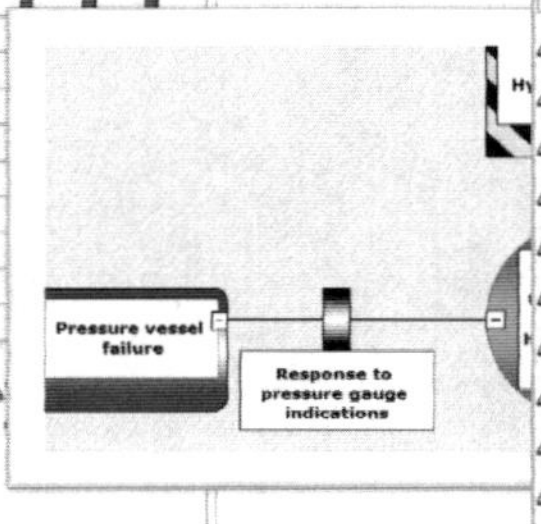

D-ZN1-G1 Unit 010 Zone 1	Off (+7/-6)	No more than 2 su
4431A Toxic Gas Detector	Online	2017 Aug 15 22:34
4431B Toxic Gas Detector	Off	2017
4431C Toxic Gas Detector	Online	2017
4431D Toxic Gas Detector	Online	2017
4431E Toxic Gas Detector	Off	2017
4431F Toxic Gas Detector	Off	2017
4431G Toxic Gas Detector	Online	2017
4431H Toxic Gas Detector	Off	2017
4431I Toxic Gas Detector	Online	2017
4431J Toxic Gas Detector	Off	2017
4431K Toxic Gas Detector	Online	2017
4431L Toxic Gas Detector	Off	2017
4431M Toxic Gas Detector	Online	2017

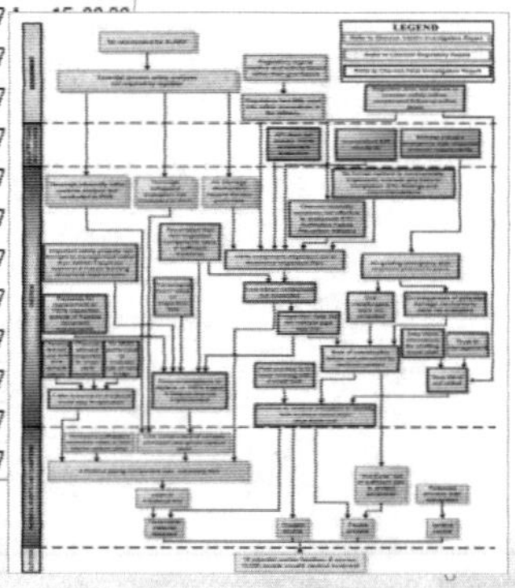

CGE
Risk Management Solutions

5-10-2017

Even more beautiful data

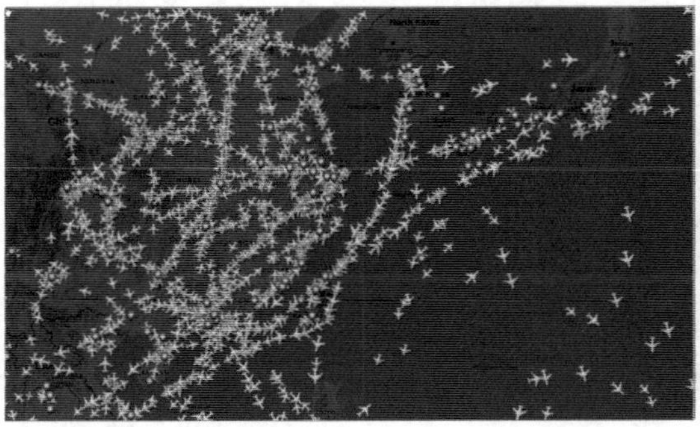

Finding a common language

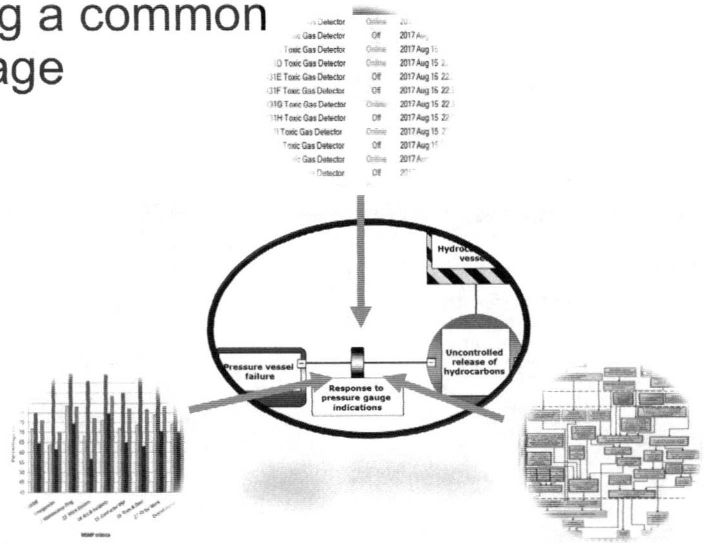

What diagrams do

Diagrams enhance cognitive processes (inference and prediction)

'Thinking with Diagrams', A. Blackwell (2001)

Station	Line(s)[*]	Local Authority	Zone(s)
Acton Town	District Piccadilly Hammersmith & City[d]	Ealing	
Aldgate East	District	Tower Hamlets	
All Saints	DLR (Stratford Branch)	Tower Hamlets	
Alperton	Piccadilly[h]	Brent	
Amersham	Metropolitan	Chiltern	
Angel	Northern	Islington	
Arnos Grove	Piccadilly	Enfield	
Arsenal	Piccadilly	Islington	
Balham	Hammersmith & City Northern	Wandsworth	3

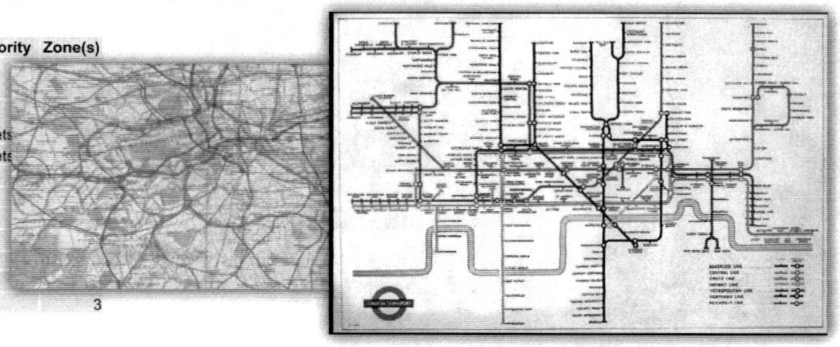

Truth – often complex and unfathomable

Models of the truth – more comprehensible

Effective models of "our view of truth"

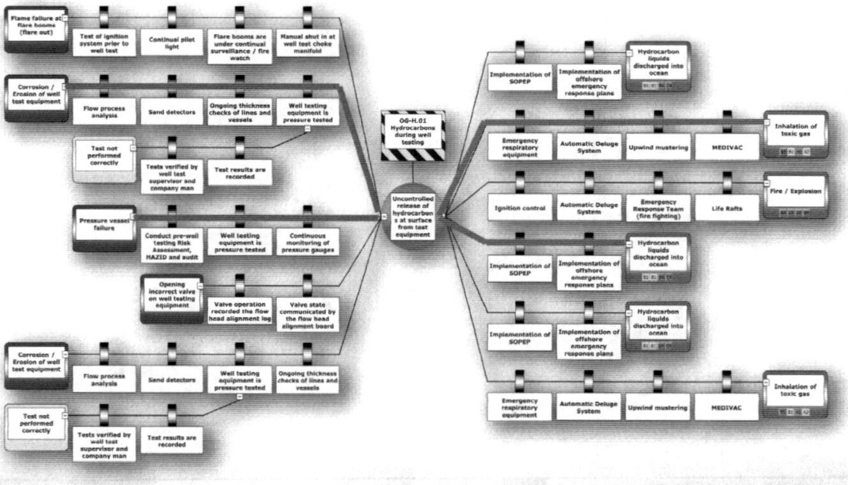

Basic uses of a Bowtie diagram

- to **Structure thinking - how things happen**
- for **Risk-based decisions and ALARP**
- for **Monitoring barrier performance / audits**
- for **Capture and share safety critical info**
- for **Risk communication up/down / training**
- for **Safety cases / Safety reports**

Proactive and Reactive

Add value Losses

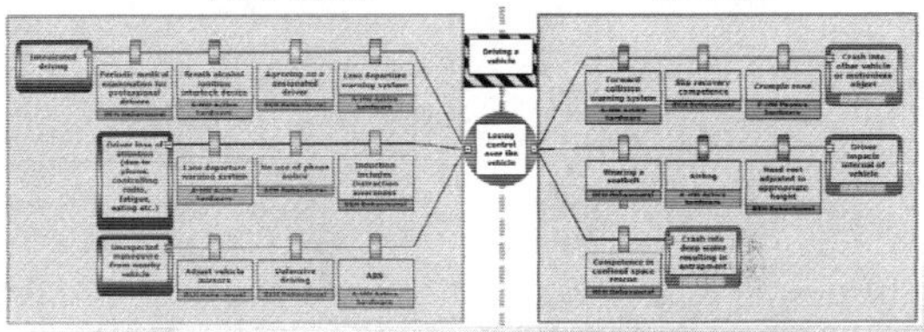

CGE
Risk Management Solutions

Incident analysis - four strategies

Incident type	Strategy
1. Simple incident	Form
2. Unique disaster	Stand-alone incident analysis
3 Serious and partly predictable	**Incident analysis /w BowTies (combi of 2 & 4)**
4. Serious but predictable	SIR (Scenario-based Incident Registration)

CGE
Risk Management Solutions

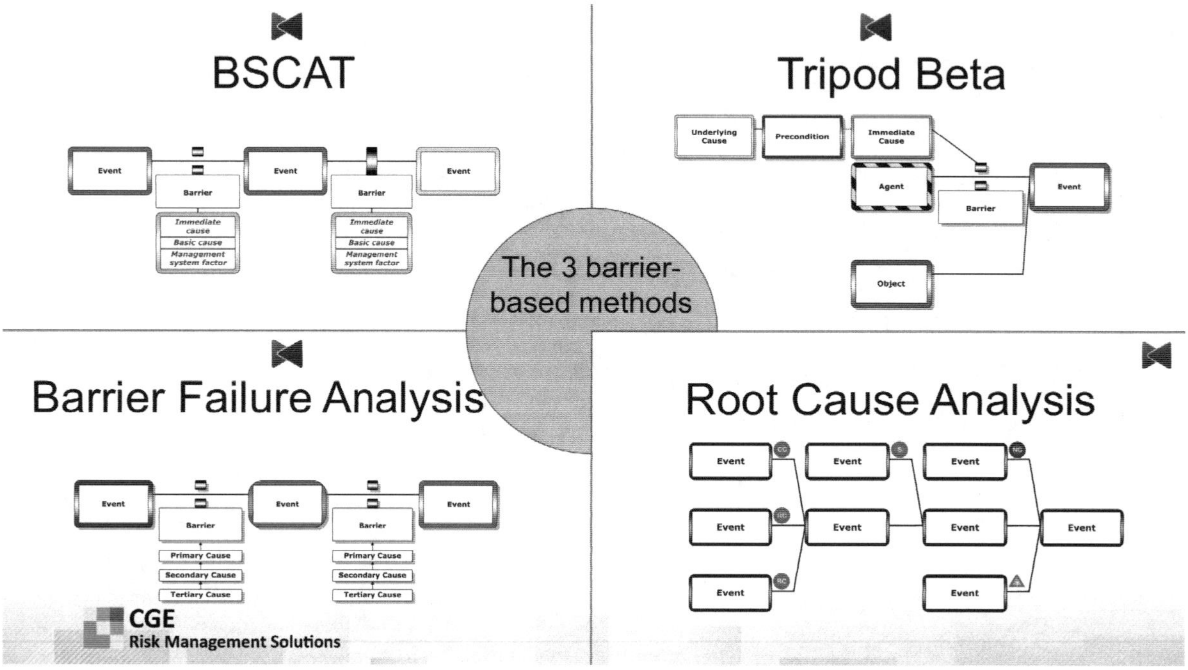

BSCAT
Event
Barrier
Immediate cause
Basic cause
Management system factor
Tripod Beta
Underlying Cause
Precondition
Immediate Cause
Agent
Barrier
Event
Object
The 3 barrier-based methods
Barrier Failure Analysis
Event
Barrier
Primary Cause
Secondary Cause
Tertiary Cause
Root Cause Analysis
Event
CGE
Risk Management Solutions

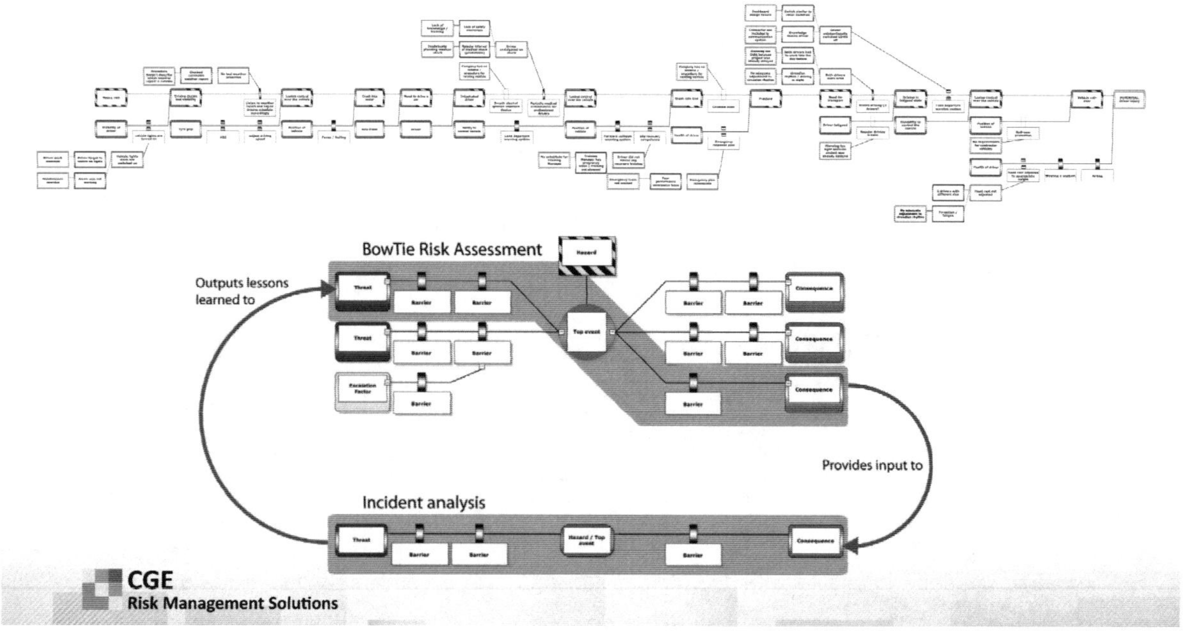

BowTie Risk Assessment
Hazard
Threat
Barrier
Top event
Consequence
Escalation Factor
Outputs lessons learned to
Provides input to
Incident analysis
Hazard / Top event
CGE
Risk Management Solutions

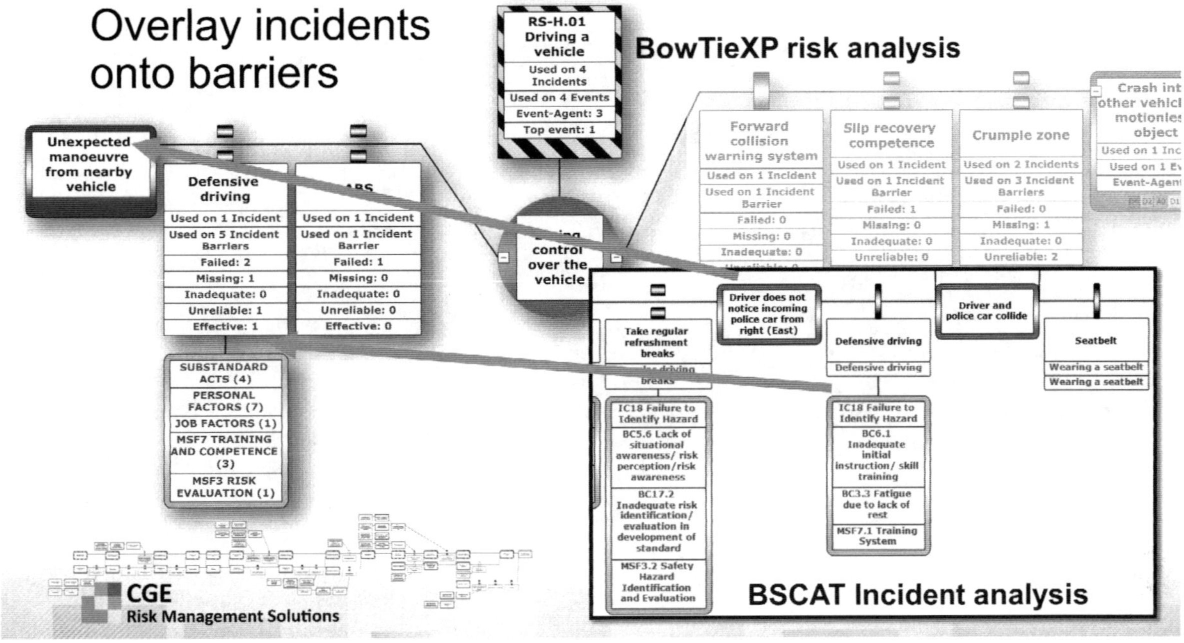

AuditXP

Surveying barriers and MS

Barrier Compliance with Standards

Audit approaches

Traditional

Barrier-based

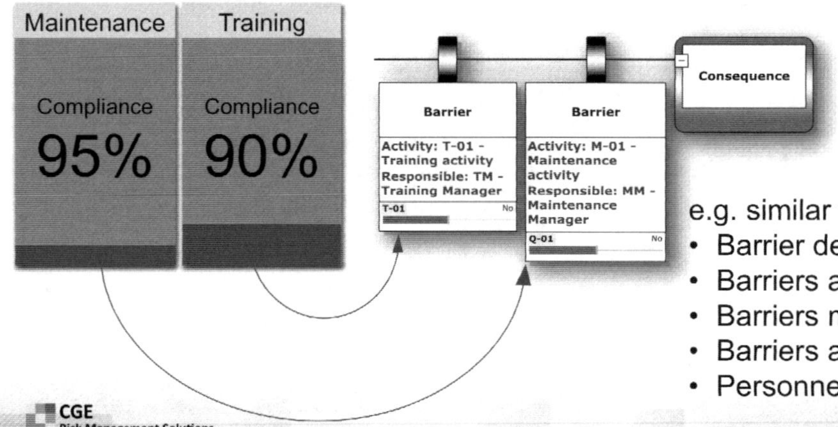

e.g. similar to deep dive audit:

- Barrier design
- Barriers are installed
- Barriers maintained & tested
- Barriers are used
- Personnel are competent

CGE
Risk Management Solutions

Questions on barriers

Simplified display option

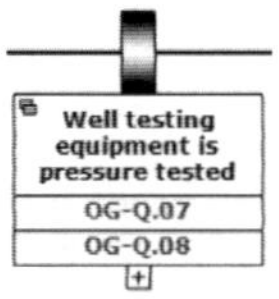

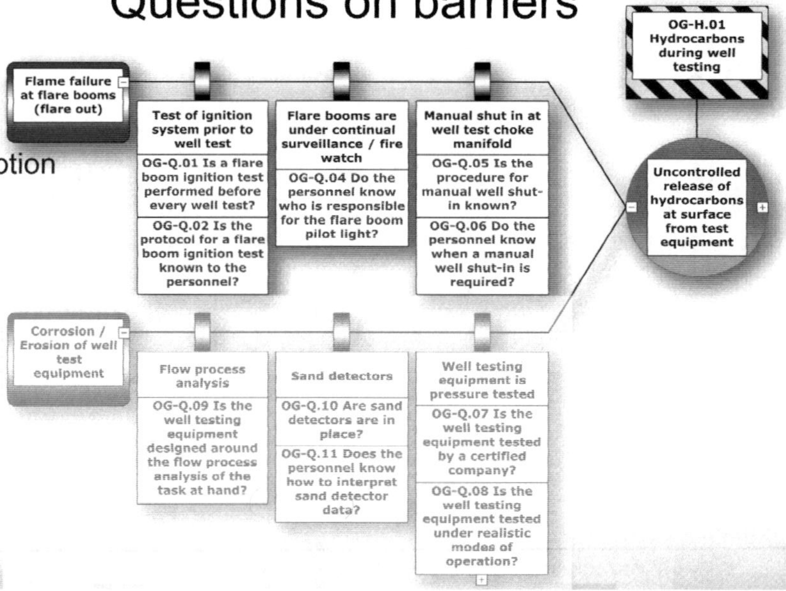

CGE
Risk Management Solutions

Barrier-based audit

- Results focus on risks, not categories
- Easy to detect weak spots
- Potential for multi-barrier impact
- MS tasks / procedures can still be audited

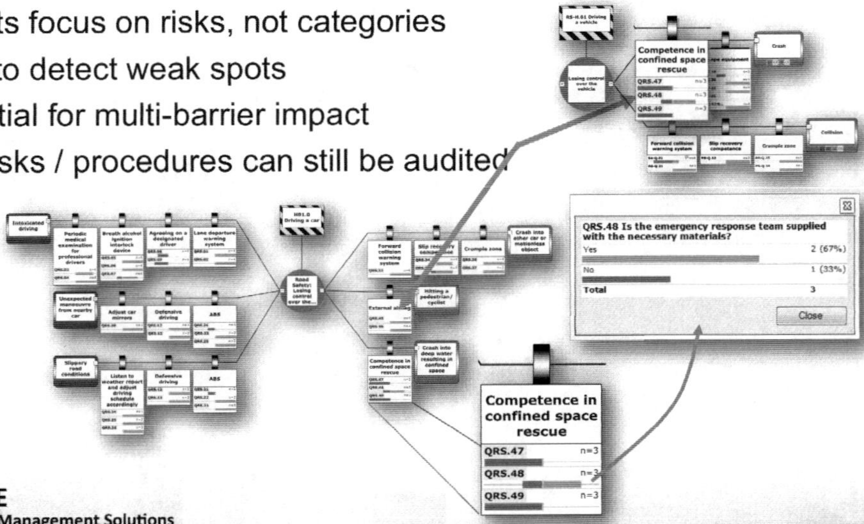

CGE
Risk Management Solutions

Equipment

Current breakdown status

Integrate maintenance backlog

MAERSK DRILLING   centrica   + more

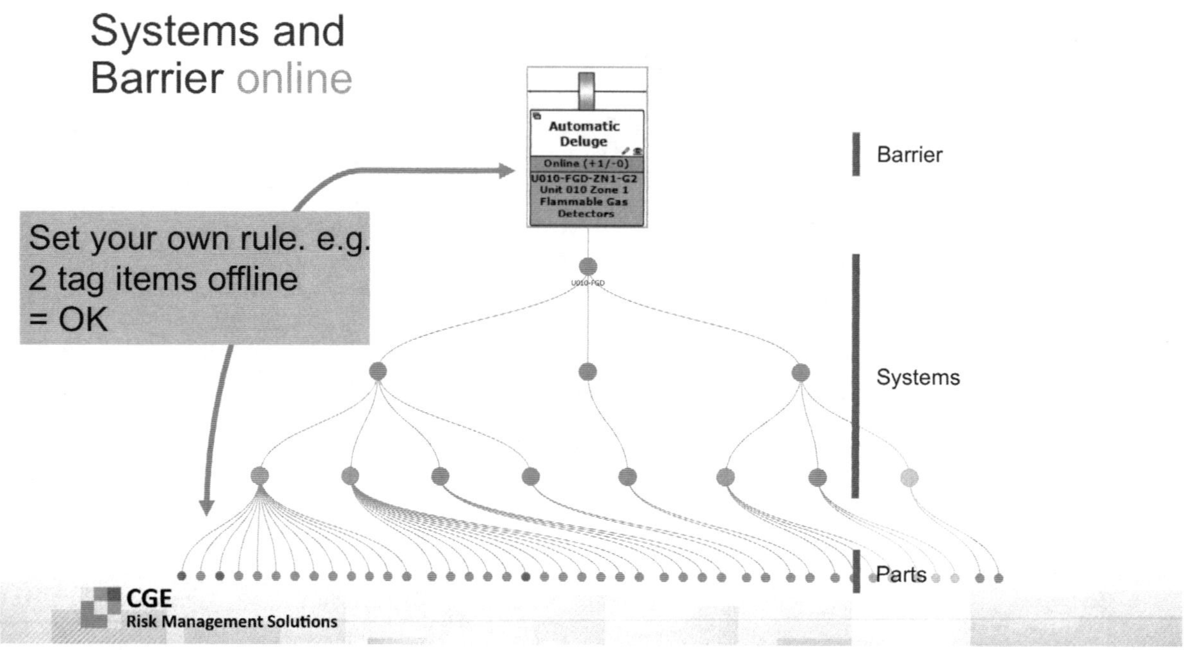
Systems and
Barrier online
Automatic
Deluge
Online (+1/-0)
U010-FGD-ZN1-G2
Unit 010 Zone 1
Flammable Gas
Detectors
Set your own rule. e.g.
2 tag items offline
= OK
U010-FGD
Barrier
Systems
Parts
CGE
Risk Management Solutions

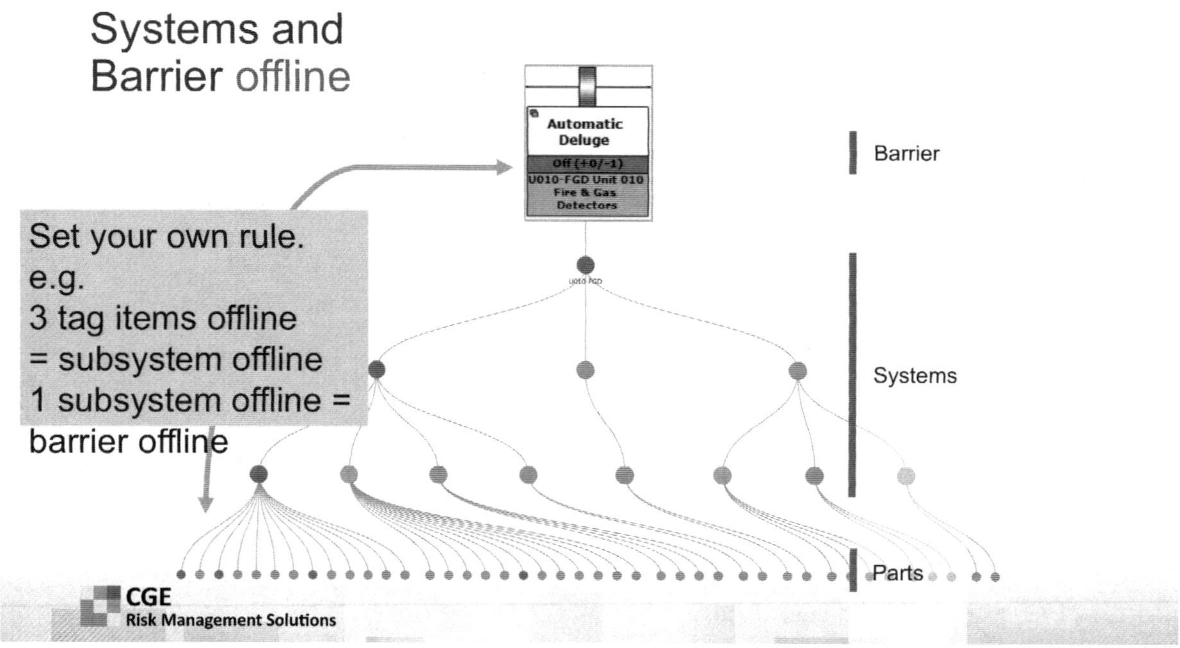
Systems and
Barrier offline
Automatic
Deluge
Off (+0/-1)
U010-FGD Unit 010
Fire & Gas
Detectors
Set your own rule.
e.g.
3 tag items offline
= subsystem offline
1 subsystem offline =
barrier offline
U010-FGD
Barrier
Systems
Parts
CGE
Risk Management Solutions

Barrier Management:

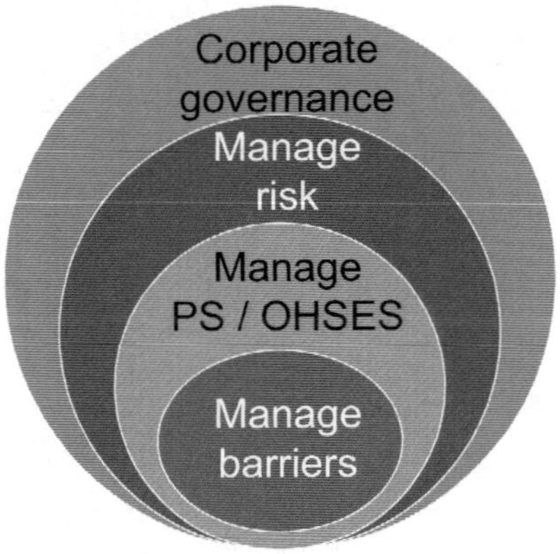

Know your barriers:

Barrier function:
The task or role of a barrier. (e.g. prevent leaks or ignition)

Barrier (system):
One or more elements.

Barrier element:
Tasks, equipment measures

Barrier Management data in BowTieXP

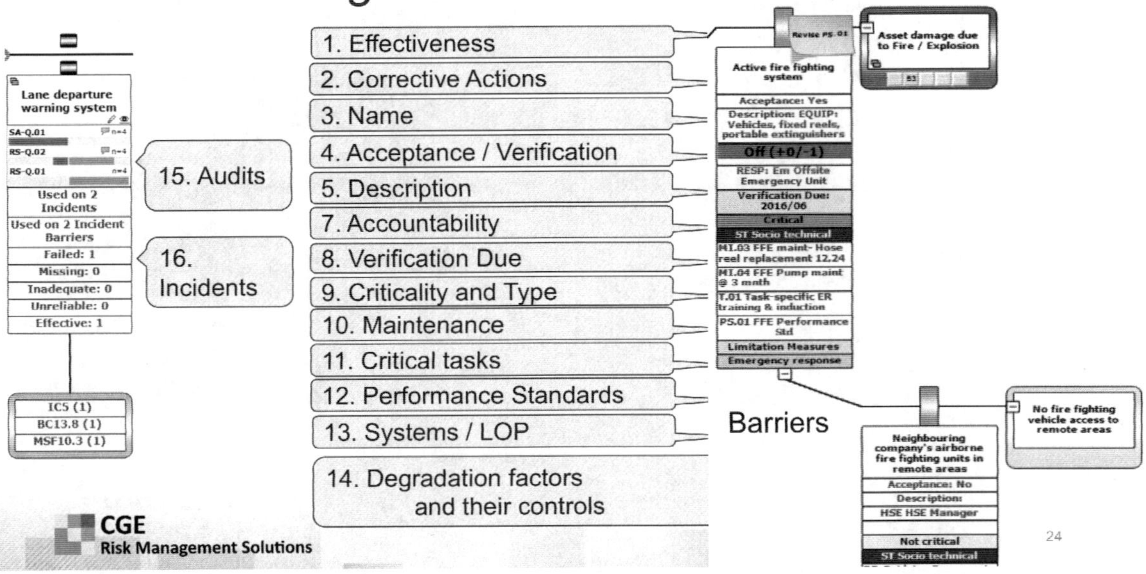

Reporting possibilities

BowTieServer BowTies Reports Audits Compliance Treatment HAZID Wiki WorkBowTie

Barrier	# Incidents	# Times	# Ineffective	State	Criticality	Effectiveness	Type	Audit
Slip recovery competence	13	21	12		Critical	Poor	Active hardware	
Use Ex equipment	11	11	10		Not Critical	Good	People	
Watchman	15	19	10		Critical	Good	Active hardware	
Ventilate, monitor and test for toxic/flammable gas	10	10	9		Not Critical	Good	Active hardware	
Weather check & avoid driving	5	7	6		Not Critical	Good	Active hardware	
Defensive driving	7	9	6		Not Critical	Good	Active hardware	
Ventilate, monitor and test oxygen level	5	5	5		Not Critical	Good	Active hardware	
Use respirator (air-supplying/air-purifying)	12	12	5		Not Critical	Very Poor	Continuous hardware	
Use full chest harness + lifeline	7	7	4		Not Critical	Good	Process	

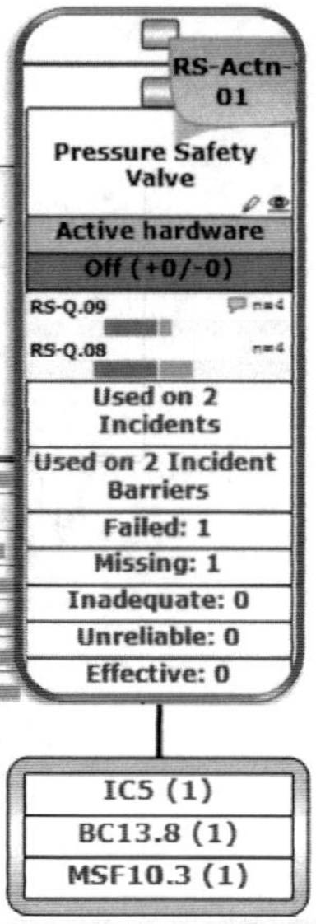

CGE
Risk Management Solutions

Thank you

Paul Haydock
p.haydock@cgerisk.com
www.cgerisk.com

CGE
Risk Management Solutions
Making risk understandable

CCPS – Publishing next quarter

(Not actual book appearance)

High Walled Diked Tanks | AIChE Academy

- Conference: CCPS Middle East Conference on Process Safety
- Year: 2017
- Time:
  Tuesday, October 10, 2017 - 2:30pm-3:00pm

Authors:
Mathkari, N., Sadara Chemical Company

Best Practice: Secondary containment of Flammable storage

Risks involved in storage of flammables in tanks are well known to industry.

This paper will discuss one of the best practices to manage this risk.

Low pressure tanks, per API 650 or API 620 are often used for storing hydrocarbons. When it comes to secondary containment, following options are available.

1. Regular common diking around multiple tanks
2. Sloping with a remote impoundment
3. Double walled tanks
4. **4. High walled dikes.**

High walled dike design makes use of a concrete wall around the tank going to the same height as tank forming an annular space that is left open at the top. The annular ring space between the dike and tank wall serves as full spill containment volume for the primary product tank. High walled design stands out as the best practice offering several advantages as listed below.

1. Concrete wall also acts as a fire wall minimizing impact to surrounding property
2. Protecting the tank from surrounding pool fires
3. Reduce need for diked area, separation distance from other assets and hence real estate
4. Reduce surface area from dike and hence lower evaporation rates. Reduced vapor cloud has favorable impact on risk from flammable or toxic chemicals
5. Quick detection of leak
6. Reduced need for the Fire protection foam quantity

A few downsides exist and design team must take those into account.

The Future of Geographical and Scenario Based Fire & Gas Mapping

Pascal Le Gal, Ph.D.

plegal@insightnumerics.com
www.insightnumerics.com

insightnumerics

What is Fire and Gas Mapping?

- The science of positioning flame and gas detectors.

Why is it important?

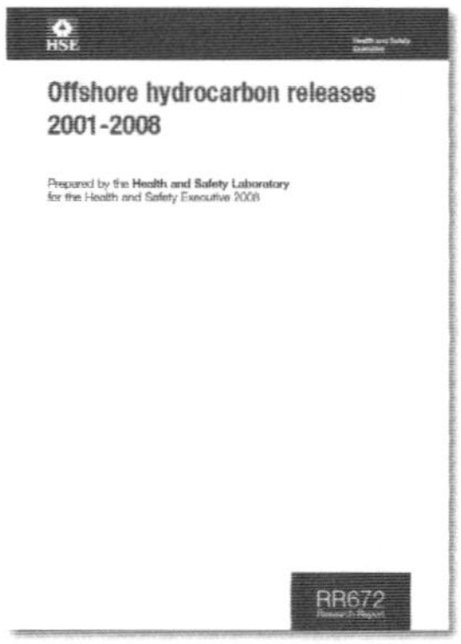
HSE

Offshore hydrocarbon releases 2001-2008

RR672

- Gas releases occur on old AND NEW platforms.
- On new platforms, the most likely source is mechanical failure, particularly flanges.
- For older platforms, it is pipe corrosion.
- ~100 major and significant gas releases per year in the UK alone.
- 1/3 of Major and 2/3 of Significant gas releases...
- ..are not detected by the gas detection systems.

Detect3D
Fire and Gas Mapping. Optimized.

Slide 2

Developed by Insight Numerics
www.insightnumerics.com

Introduction to Detect3D

- Detect3D is a modern, Windows-based software product for Fire and Gas (F&G) Mapping analysis.

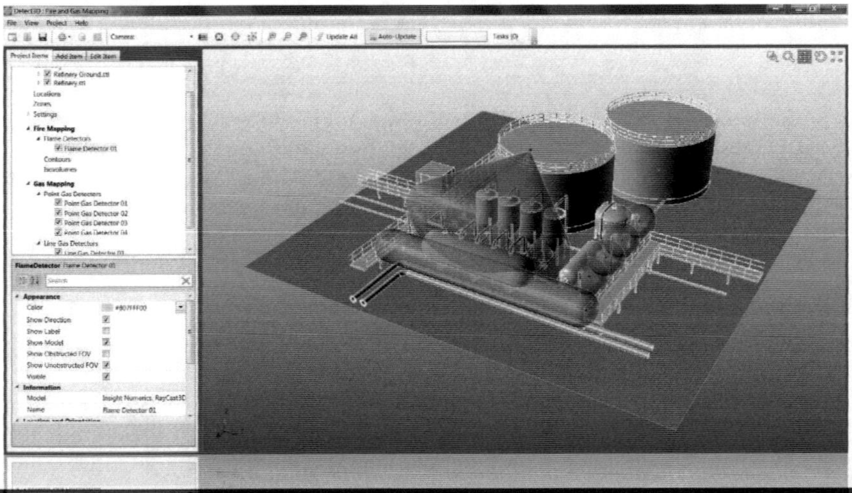

Detect3D
Fire and Gas Mapping. Optimized.

Slide 3
Developed by Insight Numerics
www.insightnumerics.com

Who Uses Detect3D?

SAIPEM · Technip · CHIYODA CORPORATION · Chevron · FLUOR · Petrofac · PERENCO · JGC · amec foster wheeler · HYUNDAI HEAVY INDUSTRIES CO., LTD. · CB&I · AkerSolutions · DNV·GL · ATKINS · ABS Consulting · SYNERGY specialist engineering · vanguard solutions · SAFETEC · BMT Fluid Mechanics · PRISM · Applus VELOSI · arc · NAPESCO INDIA · Imtech · DET-TRONICS A UTC Fire & Security Company · BLACK & VEATCH · STYLEMARK DESIGNS LTD. Draughting & Design Services · Chola MS RISK SERVICES · CORE · innovecm

Detect3D
Fire and Gas Mapping. Optimized.

Slide 4
Developed by Insight Numerics
www.insightnumerics.com

Where is Detect3D used?

- Detect3D has been used on projects worldwide for BP, Shell, Petronas, ADNOC, GASCO, Maersk, Apache and many more.

The Americas
ALPEK – FGRU Indelpro
SOFEC
Altamira

Europe
Statoil – Johan Sverdrup Platform
Maersk – Culzean Platform & FSO
Ineos – Grangemouth
Apache – multiple projects
Technip/JGC – Coral South Development Project

Asia
BP – South Caucasus Pipeline Expansion
TCO – Kazaskhstan Project
GAZPROM – Badra-Zubaida Pipeline

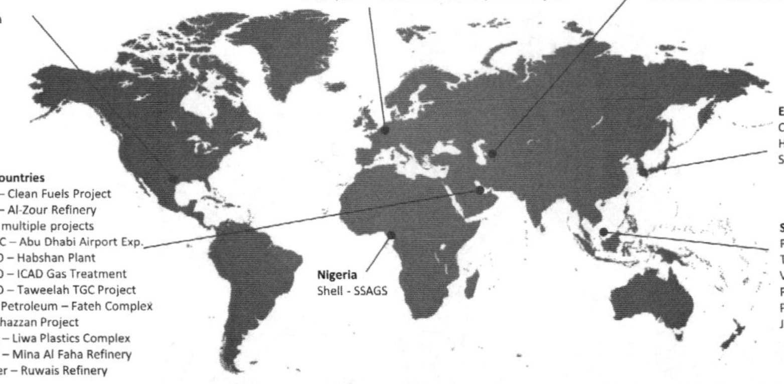

East Asia
COSCO Shipyard – FPSO
HHI – multiple projects
Seoul National University

GCC Countries
KNPC – Clean Fuels Project
KNPC – Al-Zour Refinery
KOC – multiple projects
ADNOC – Abu Dhabi Airport Exp.
GASCO – Habshan Plant
GASCO – ICAD Gas Treatment
GASCO – Taweelah TGC Project
Dubai Petroleum – Fateh Complex
BP – Khazzan Project
ORPIC – Liwa Plastics Complex
ORPIC – Mina Al Faha Refinery
Takreer – Ruwais Refinery

Nigeria
Shell - SSAGS

SE Asia
PETRONAS - FLNG 2
ThaiOil – Clean Fuels Project
Vestigo - MaMPU FPSO
Pertamina – PPP
Premier Oil - Natuna Sea Block A
JX Nippon – Helang Project

Detect3D
Fire and Gas Mapping. Optimized.

Slide 5
Developed by Insight Numerics
www.insightnumerics.com

The Past...

- Fire and Gas Mapping, when performed, used simple 2D techniques and software.

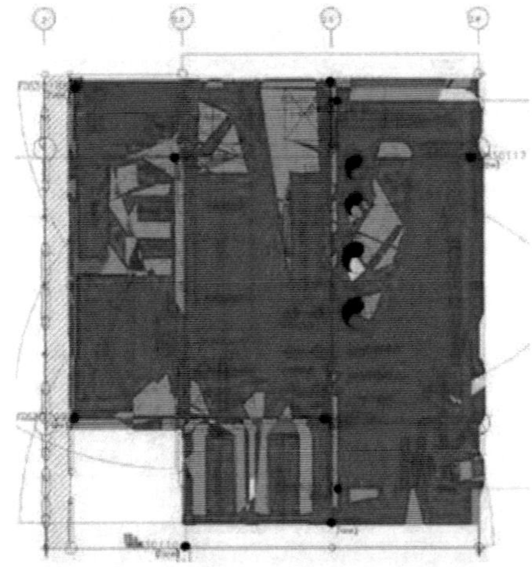

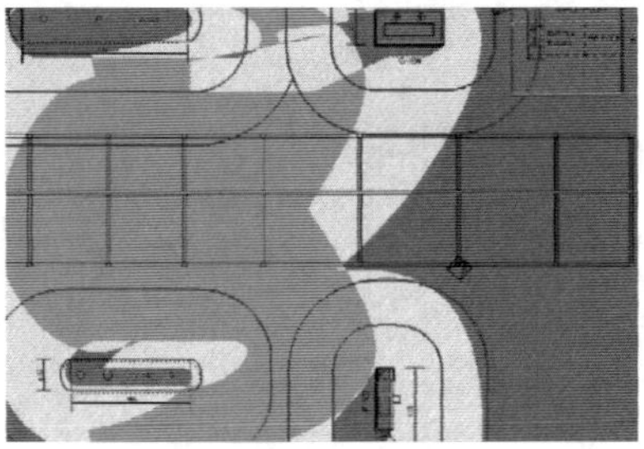

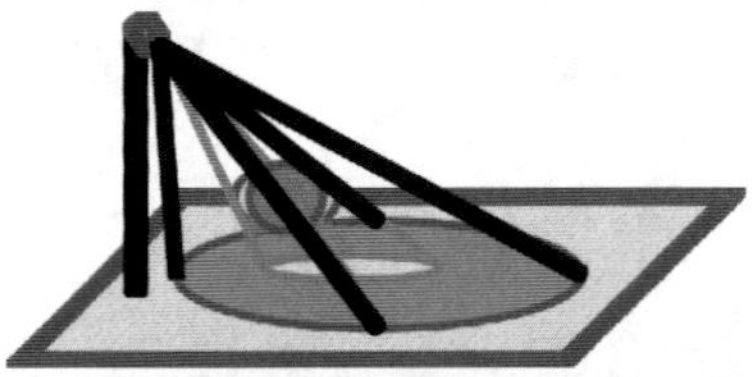

Detect3D
Fire and Gas Mapping. Optimized.

Slide 6
Developed by Insight Numerics
www.insightnumerics.com

The Present...

- Integration with 3D CAD

Above Screenshot of Detect3D showing a typical CAD file for an onshore site.

Right Further examples of CAD used in Detect3D for 3D F&G Mapping projects.

Detect3D
Fire and Gas Mapping. Optimized.

Slide 7

Developed by Insight Numerics
www.insightnumerics.com

The Present...

- Fire mapping uses ray casting to accurately calculate the detectors Field of View...

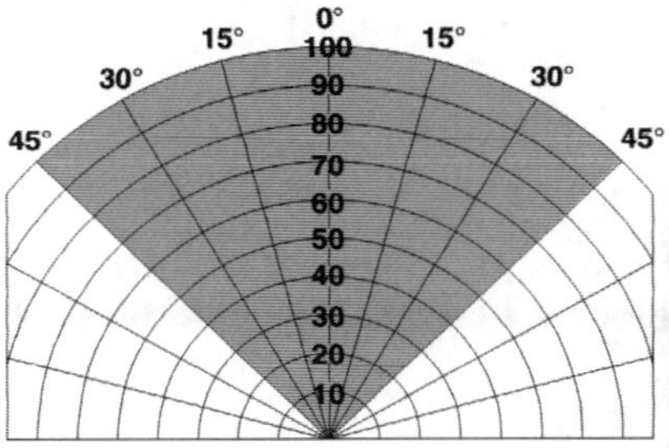

Above
Manufacturer's data for a flame detector FOV

Right
An obstructed FOV (green) calculated by Detect3D

Detect3D
Fire and Gas Mapping. Optimized.

Slide 8

Developed by Insight Numerics
www.insightnumerics.com

The Present...

- ...to arrive at coverage statistics and visualization.

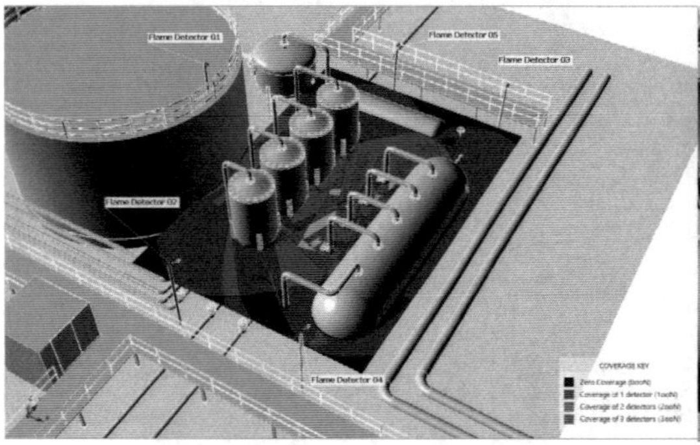

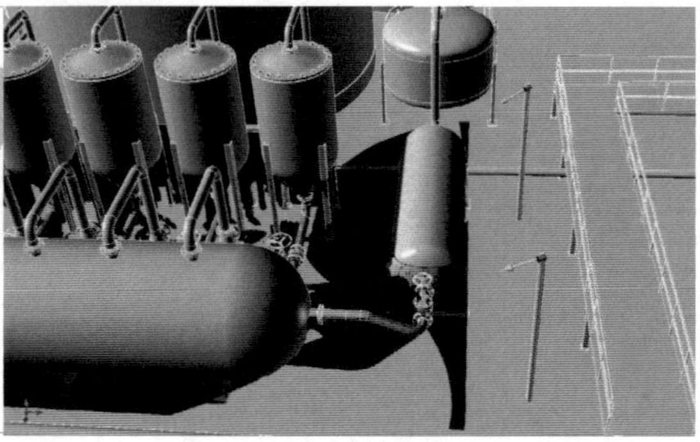

Left
Coverage contour at 2 meters above ground level.

Right
Zero Coverage Isovolume

Detect3D
Fire and Gas Mapping. Optimized.

Slide 9
Developed by Insight Numerics
www.insightnumerics.com

The Present...

- The coverage of gas detector layouts is calculated based on the "design gas cloud" approach popularized by the UK HSE, BP and Shell.

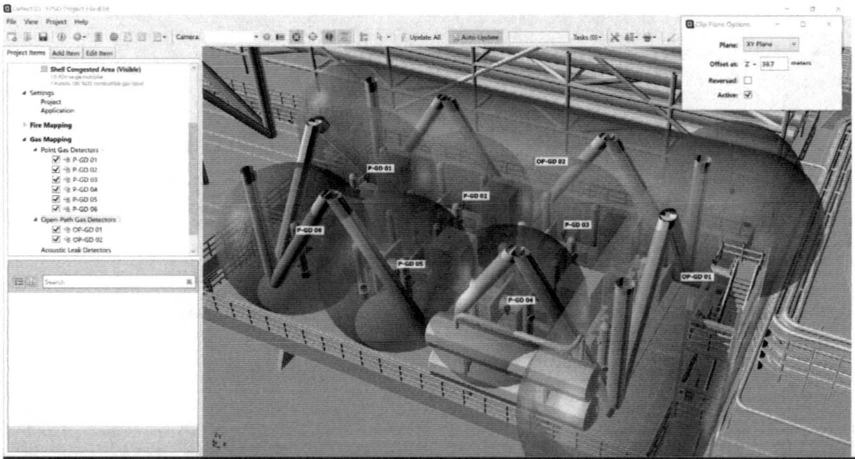

Left
Point gas & open path gas detectors highlighted on a FPSO deck. The blue spheres indicate the "field of influence" of the detector, depending the gas cloud size required for the coverage assessment.

Multiple gas cloud sizes can be used for the gas mapping evaluation.

Cut plane view

Detect3D
Fire and Gas Mapping. Optimized.

Slide 10
Developed by Insight Numerics
www.insightnumerics.com

Room for Improvement?

- The geographic method:
  - Provides acceptable results for F&G Mapping based on years of experience and research.
  - Is accepted by many authorities.
  - Does not account for any physical effects such as wind speed, with direction, gas cloud, fire.
- Why not replace the geographical method with scenario based modeling that account for all physical effects?
  - 8 wind directions, 3 wind speeds
  - 50 leak locations
  - 4 leak directions, 3 hole sizes
  - 8 x 3 x 50 x 4 x 3 = 14,400 scenarios!

Detect3D
Fire and Gas Mapping. Optimized.

Slide 11

Developed by Insight Numerics
www.insightnumerics.com

Scenario based methods

- We have >10,000 scenarios per module – can we use simple dispersion models?
  - Only part of the physical environment is included.
  - Is an inaccurate scenario based analysis better than a geographical assessment?
- Let's use CFD!
  - Costs are high – just reroute the cost to more detectors?
  - No established methodology.
  - No industry testing or JIP to support a full CFD method.
  - Higher analysis costs…but more detectors!

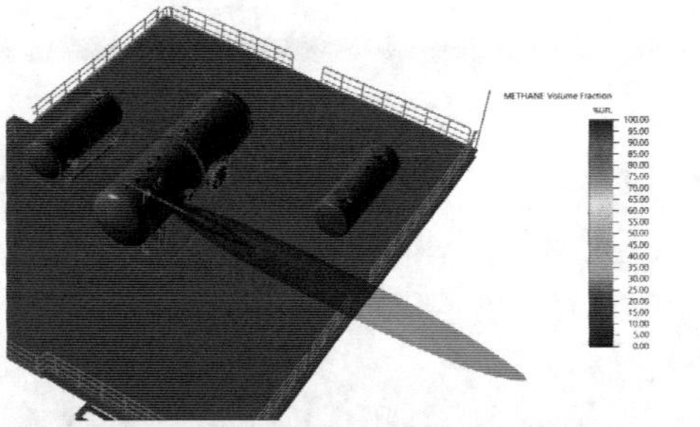

Detect3D
Fire and Gas Mapping. Optimized.

Slide 12

Developed by Insight Numerics
www.insightnumerics.com

New CFD software tool

in:Flux
INTELLIGENT CFD SOFTWARE

- Initially for ventilation and dispersion
- Fire and explosion modelling to follow-up in 2018

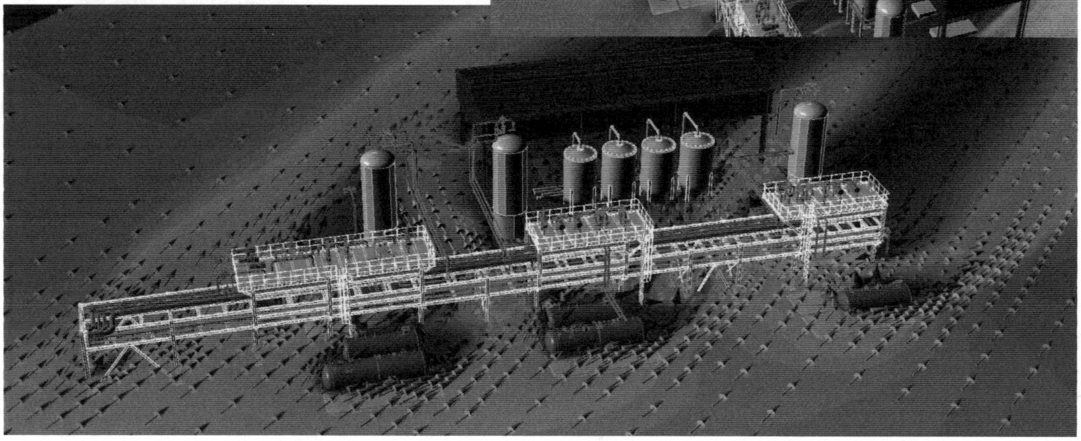

CFD – a way forward

- Detect3D will have a CFD capability from in:Flux (2018)
- In:Flux to be released formally by November 2017

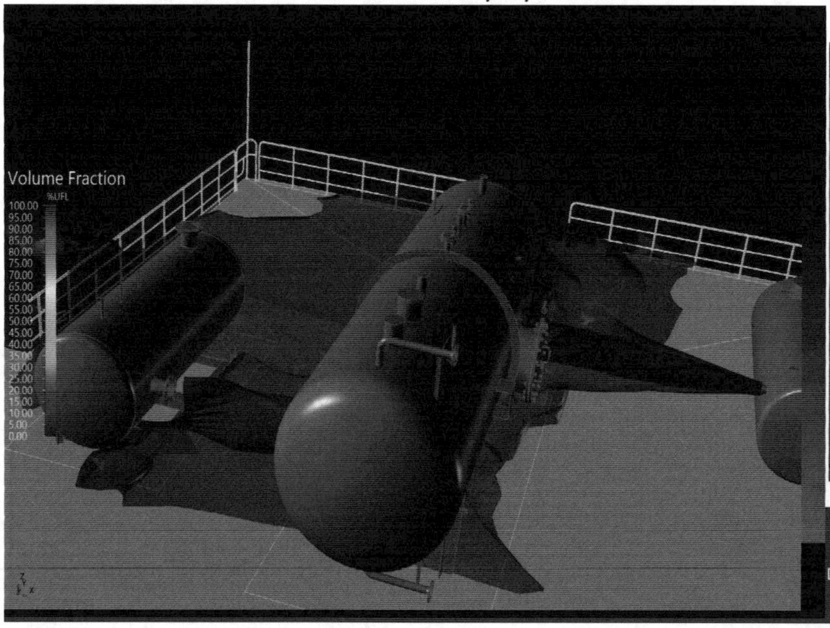

Slide 14

Developed by Insight Numerics
www.insightnumerics.com

CFD – a way forward

- The most likely way forward is to use a combination of geographical and scenario based modeling, based on the strengths of each method.
- Use geographic method in highly congested environments.
- Use scenario based methods to address weaknesses of the geographic method, particularly:
  - Detection at a distance, e.g. H_2S / toxic gas detection using open-path detectors at a site perimeter.
  - Assess detection time.
  - Calculate detector alarm settings.
  - Better assessment of technologies – point vs open path vs acoustic.

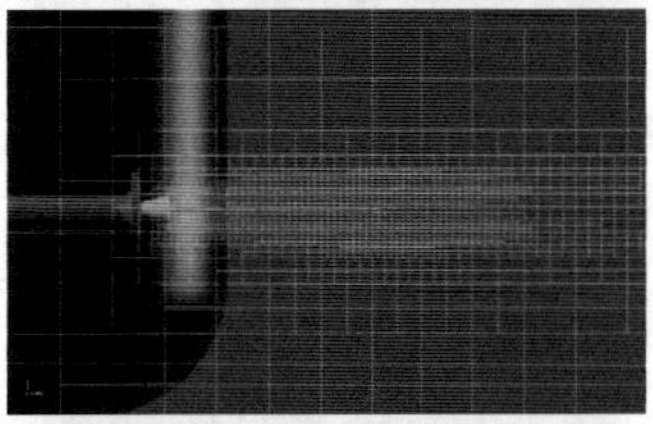

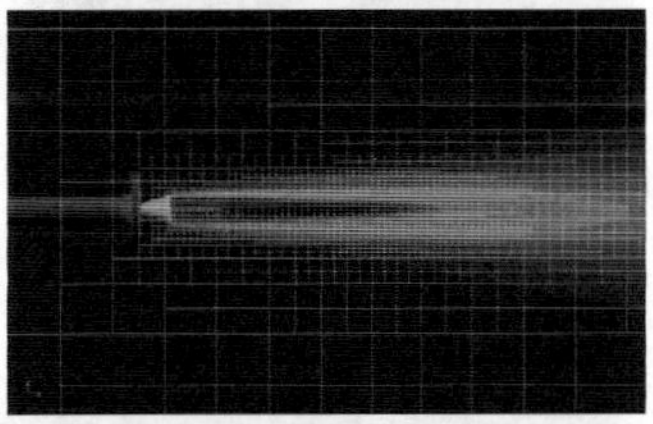

The Future

- Progress in Fire and Gas Mapping technology will be focused on CFD simulations of gas and fire scenarios.
- There are many unanswered questions:
  - How many scenarios are necessary?
  - How can we validate the layout?
  - What is the relative detector count for scenario and geographic coverage?
  - Is CFD feasible, or necessary, for flame detection mapping?
- We need a joint industry effort to research these questions, involving oil majors, regulators, universities, consultancies and software companies.
- The technology is developing fast and the industry needs to keep pace.

Questions?

Pascal Le Gal, Ph.D.
Tel: +971 50 666 7935
plegal@insightnumerics.com
www.insightnumerics.com

Detect3D
Fire and Gas Mapping. Optimized.

in:Flux
INTELLIGENT CFD SOFTWARE

Slide 17

Developed by Insight Numerics
www.insightnumerics.com

ABS Group

MIDDLE EAST PROCESS SAFETY CONFERENCE | 9 - 11 OCTOBER 2017 Bahrain International Exhibition and Convention Centre

CCPS MIDDLE EAST PROCESS SAFETY CONFERENCE

Achieve Quick Improvements in Process Safety

Steve Arendt, PE
VP-Global Oil, Gas & Chemicals
+1-281-673-2914
sarendt@absconsulting.com
www.abs-group.com

Steve Arendt, P.E., Vice President - Global Oil, Gas, and Chemicals

- 40 years in process safety and risk assessment
- Conducted 100s of PSM system design/implementations, audits, incident investigations, HAZID/HAZOP/LOPA/QRAs, best practice reviews, and safety culture evaluations
- ABS Group manager for the BP-Baker Panel PSM reviews
- 80+ articles and books on PSM and risk management, including CCPS Guidelines on
  - **Risk Based Process Safety**
  - **Implementing PSM, 2nd edition**
  - **Management of Change**
  - **Guidelines for Hazard Evaluation Procedures, 2nd Edition**
- CCPS Fellow and recipient of MKOPSC Merit Award

ABS Group

2

Discussion Outline

- Background
- Foundations of effective PSM
- Quick win tactics for PS improvement
- Preserving/resurrecting your “drive to ZERO”
- Closing suggestions

ABS Group

3

MIDDLE EAST PROCESS SAFETY CONFERENCE | 9 - 11 OCTOBER 2017

Background

- Many companies have adopted continuous improvement as their primary strategy toward pursuing Operational Excellence in Process Safety
- CCPS has extended this strategy with its Vision 2020
- However, some companies (or countries) may not be succeeding to strive for the highest levels of process safety performance
- Perhaps, they need to “step-back” and re-implement some basic steps to create a solid foundation for moving forward in process safety

ABS Group

4

MIDDLE EAST PROCESS SAFETY CONFERENCE | 9 - 11 OCTOBER 2017

Foundation for an Effective PSM System

- Optimize your PSM/HSE system based upon hazards, complexity, resources and culture using the CCPS ***Guidelines for Risk Based Process Safety***
- Use the essential features toolbox in RBPS for a fit-for-purpose" PSM system
- Avoid well-know PSM implementation "failure modes"
- Secure your foundation with competency and leadership at all levels

ABS Group

MIDDLE EAST PROCESS SAFETY CONFERENCE | 9 - 11 OCTOBER 2017

5

HSE/Process Safety Culture and Leadership

- **Culture** is the tendency in all of us – and our organization – to want to do the **right thing** in the **right way** at the **right time**, **ALL the time** – even when no one is looking
- **Leadership** – throughout the organization - is critical for a good culture
- Best if you assess HSE/PS culture using interconnected methods/data sources:
  - Interviews
  - Surveys
  - Small group workshops and focus groups
  - Site observations and discussions
  - Performance data analysis

ABS Group

MIDDLE EAST PROCESS SAFETY CONFERENCE | 9 - 11 OCTOBER 2017

6

Create Quick/Inexpensive Process Safety Improvements

- Implement a management review process
- Improve process safety leadership at the shift leader level
- Create, monitor, and react to simple process safety metrics
- Make everyone an MOC sensor
- Do simple, effective root cause analysis to learn from everything
- Conduct effective process safety reviews/audits

ABS Group

MIDDLE EAST PROCESS SAFETY CONFERENCE | 9 - 11 OCTOBER 2017

7

Management Review Essential Features

- Written management practice procedure
- Top site leader
- Scheduled, not a surprise
- Special, not embedded
- Sufficient time
- Expect preparations
- Regularly conducted
  - Sites - monthly to quarterly; division/company - quarterly to annually
- Agenda and pre-read sent out in advance
- Attendance requirements
  - Site leader, SLT, element leader, worker, contractor

ABS Group

MIDDLE EAST PROCESS SAFETY CONFERENCE | 9 - 11 OCTOBER 2017

8

Management Review Essential Features

- Cover all PS/HSE elements are addressed during 1/2 audit-cycle – adjust frequency for "needy" elements
- Typical agenda - common to all meetings
  - Performance
  - Audits
  - Incidents
  - Metrics
  - Resources/headcount - What is needed to pursue improvement
  - Future challenges
- Typical questions/topics for each element
- Action item follow-up from previous MR
- Documented results and actions

ABS Group

9

MIDDLE EAST PROCESS SAFETY CONFERENCE | 9 - 11 OCTOBER 2017

Leadership Site Engagement Essential Features

- **Leadership Site Engagement Preparation**
  - Scope
  - Collect Information
  - Review of Major Hazards
  - Set Objectives
  - Personal Preparations
  - Complete Plan
- **Leadership Site Engagement Execution**
  - Site orientation
  - Site Leadership Meeting
  - Contractor Leadership Meeting
  - Workforce Meeting
  - Site Walk Around
- **Leadership Site Engagement Closure**
  - Workforce and Contractor Leadership Meetings
  - Site Leadership Meeting
  - Complete Closure

Intentional design of engagement content

Predetermined questions and areas to review

ABS Group

10

MIDDLE EAST PROCESS SAFETY CONFERENCE | 9 - 11 OCTOBER 2017

Improve PS/HSE Leadership at the Shift Leader Level

Equip everyone with

Advanced PS/HSE supervision skills
+
PS/HSE ownership
+
Soft skills
+
Culture skills
=

Barrier Guardians and Culture Agents©

ABS Group

11

MIDDLE EAST PROCESS SAFETY CONFERENCE | 9 - 11 OCTOBER 2017

Create, Monitor, and React to Simple PS Metrics

- If you could only have a few, which ones would you pick?
  1. Process Safety incidents – ANSI/API RP 754 Tiers 1 and 2
  2. Process Safety incident precursors – RP 754 Tier 3
  3. Failure to follow procedures/SWPs – BBS at-risk observation or SWP audit rate
  4. Failure to fix identified process safety problems – action item backlogs or aging, equipment deficiencies backlogs
  5. Failure to identify process safety deficiencies– inspection (all sources) backlog, failure to identify/report incidents or do adequate RCA
  6. Failure to assess risk – MOC circumvention or low quality, PHA schedule backlog, PHA quality review
  7. Safety culture weaknesses – Map RCs of incidents to cultural causal factors

ABS Group

12

MIDDLE EAST PROCESS SAFETY CONFERENCE | 9 - 11 OCTOBER 2017

Choose Your Metrics Wisely...

- Address all pyramid levels?
- Includes most risk barriers?
- Indicators have accident prevention value?
- Reproducible/consistent?
- Auditable?
- Prevention power/diagnostic?
- Sensitive/timely?
- Easy to use/not too expensive?

ABS Group

13

MIDDLE EAST PROCESS SAFETY CONFERENCE | 9 - 11 OCTOBER 2017

Metrics Advice

- Don't pick too many
- Make sure they roll up properly
- Make sure they add value
- Don't just pick things you can measure; make certain they affect accident risk
- Think through how you will use them; anticipate unintended behaviors
- Make them visible – positive culture influence
- Don't be afraid to change them

ABS Group

14

MIDDLE EAST PROCESS SAFETY CONFERENCE | 9 - 11 OCTOBER 2017

Getting Value from Metrics

- Making them up alone doesn't do anything
- Proper refresh rate, consistency/quality in calculations will generate credibility
- To get 1st-order benefits, make sure they filter through the chain of command to affect the performance of departments and people
- To get ancillary benefits, you can't hide them. You must widely advertise them and talk about them to get the culture improvement benefit
- Once you see there is an issue, you MUST act. Watching metrics alone doesn't affect performance

ABS Group

15

Make Everyone an MOC Sensor

- MOC problems is a leading cause of process safety incidents
- Many MOC programs are becoming a collage of electronic jig-saw puzzle pieces that is hard to keep knit together Monitor three types of MOC leading indicators:
  - Circumvention of MOC process (regular scrutiny of CCMS work order logs and nurturing of maintenance planners will help avoid this one
  - Quality deficiencies in MOC reviews becoming endemic due to electronic system "tick-boxing"
  - Broader awareness of EVERYONE that could be affected using more effective delivery and follow-up schemes

ABS Group

16

Apply Effective Root Thinking EVERYWHERE

- Do simple, effective root cause analysis to learn from everything
- Prepare for using more than one RCA method, based upon need
- Train your investigators, but you MUST exercise them or quality will suffer
- Apply RCA for chronic problems – repeat PHA action items and audit non-conformancies
- And develop SMARTER action items with an assurance process

ABS Group

17

Action Items Should Be SMARTER

- **S**pecific – Action must be specific. If it states an objective to be accomplished, it should describe at least on way this objective might be accomplished. It cannot be generic in nature. There should not be multiple actions hidden in the action.
- **M**easurable – It should be possible to measure when the action is complete
- **A**ccountable – The action must be assigned to a responsible person and with a target date for completion
- **R**elevant – The action must address the causes identified through improved prevention, detection, and/or correction of the causes
- **T**imely – The target date for completion must be timely and realistic
- **E**valuated – Criteria for evaluating recommendation effectiveness are identified
- **R**eviewed – The completion and effectiveness of the action are reviewed (verified)

ABS Group

18

Conduct Effective Process Safety Reviews/Audits

- If you have effective metrics program combined with an active management review process, you'll be able to extend out the audit cycle
- And when audits are needed, your metrics program can help target elements that need the scrutiny, thus simplifying audits

ABS Group

19

MIDDLE EAST PROCESS SAFETY CONFERENCE | 9 - 11 OCTOBER 2017

For Others, How Do You Preserve Your "Drive to ZERO"

- When improvement stops, what do you do?
- People wonder why, and there is pressure from many directions – internal and external
- Key questions
  - Can you believe your measuring data and methods?
  - If so, can you improve?
  - Then, how can you sustain it?
  - How to continue driving to zero?

ABS Group

20

MIDDLE EAST PROCESS SAFETY CONFERENCE | 9 - 11 OCTOBER 2017

Diagnosing HSE/PS Performance Deficiencies

- Examine learning mechanisms and corrective action processes
  - Incident investigations and root cause analysis effectiveness
  - Audit effectiveness
  - Action item completion work processes and results
- Examine leading indicators to see if they have HSE/PS improvement value vs just things easy to collect and are really being used to drive performance
- Examine the effectiveness of existing behavior based safety (BBS) program – many BBS programs lose value and need to be re-energized
- Do an HSE/PS culture disease screening – determine whether there is evidence of chronic problems that never stay fixed
- Conduct an HSE/PS culture evaluation
- Then, improve the areas where the problems are

ABS Group

21

Safety Culture Disease Screening

Do you have any of the following symptoms?

- Chronic work backlogs
- Problems that never seem to get better
- Poor reporting
- Investigations identify symptoms, not root causes
- Many incidents involve "people not following procedures"
- Repetitive barrier degradation patterns
- Repeated root causes – over and over and over…
- Corrective actions don't address root causes
- Fixes don't stay fixed

ABS Group

22

Culture Degradation Progression

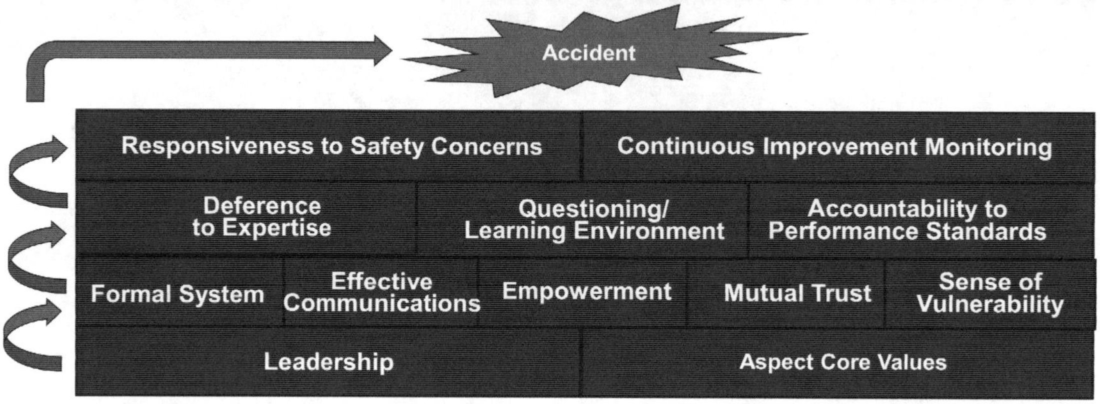

ABS Group

23

MIDDLE EAST PROCESS SAFETY CONFERENCE

Foundations on the Pyramid

Pursue Culture from BOTH Directions

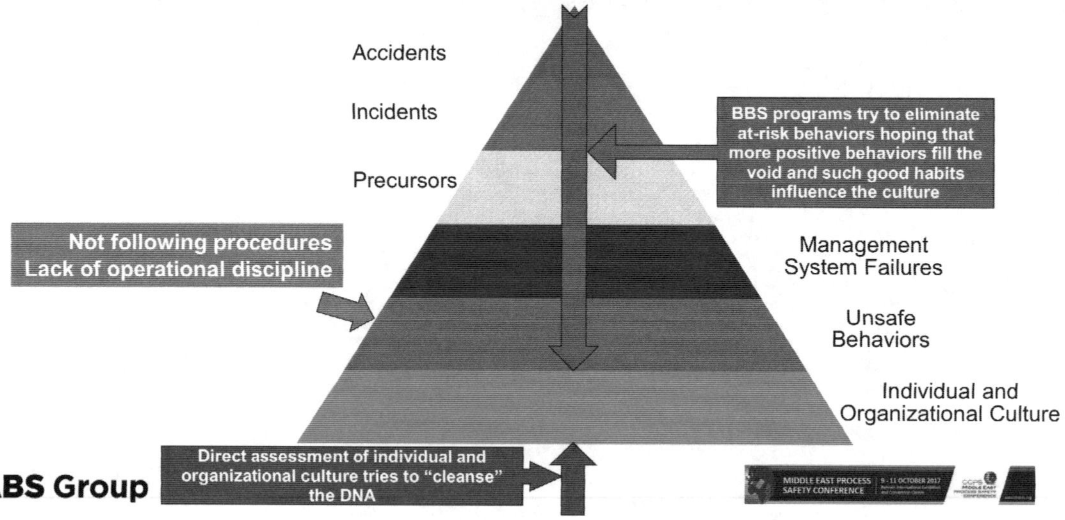

ABS Group

MIDDLE EAST PROCESS SAFETY CONFERENCE

Closing Suggestions

- Use RBPS to help "manage generational change" in Process Safety
- Make everyone a **Barrier Guardian and Culture Agent©**
- Monitor "values drift" as execution occurs down the chain of command
- Monitor for signs of safety culture disease
- Vaccinate influencers with PS Leadership

ABS Group

MIDDLE EAST PROCESS SAFETY CONFERENCE 9-11 OCTOBER 2017

Continuous, Sustainable Improvement in HSE/Process Safety Performance Demands...

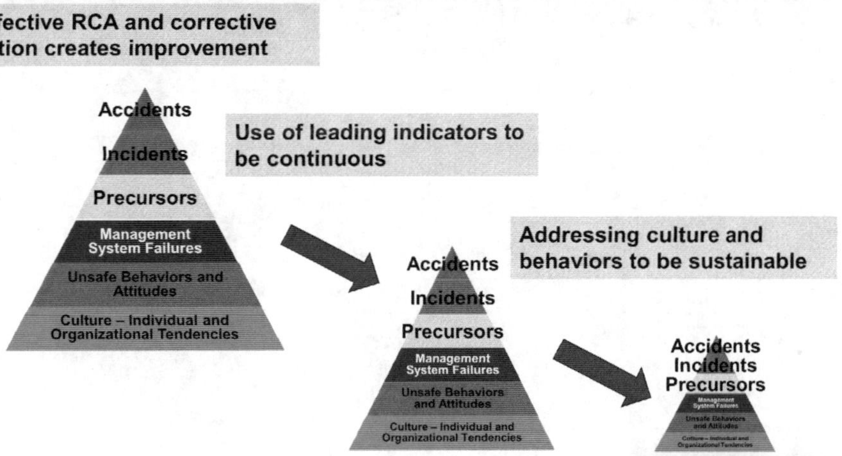

ABS Group

MIDDLE EAST PROCESS SAFETY CONFERENCE 9-11 OCTOBER 2017

ABS Group
27

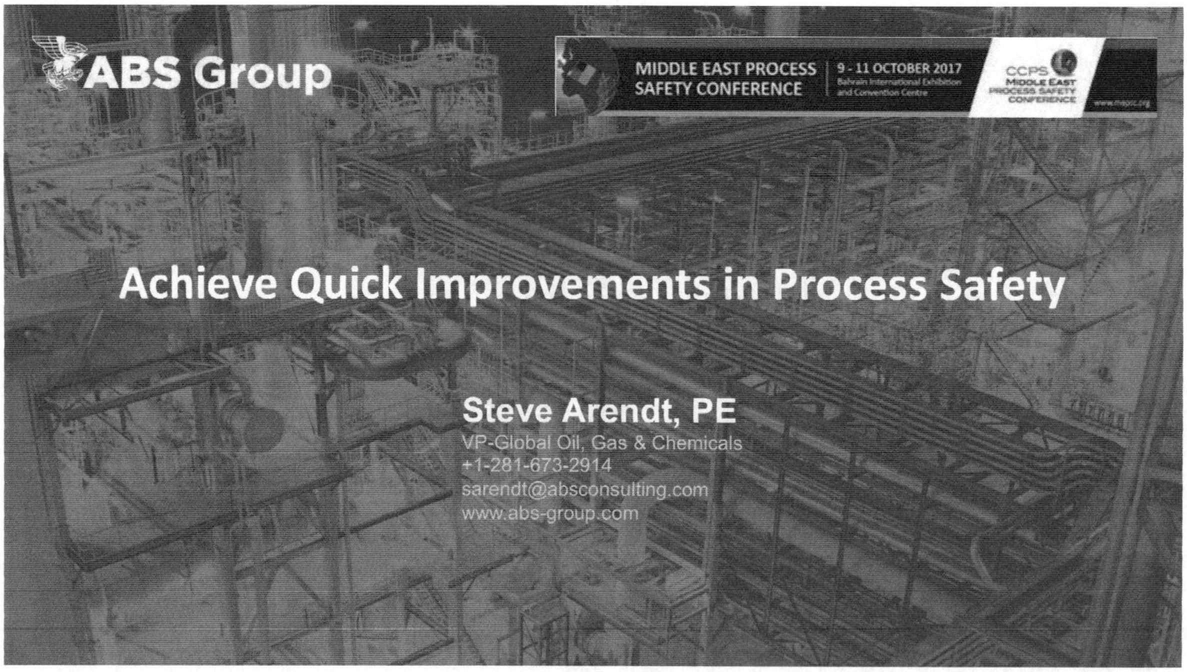
ABS Group
MIDDLE EAST PROCESS SAFETY CONFERENCE
9 - 11 OCTOBER 2017
Bahrain International Exhibition and Convention Centre
CCPS
MIDDLE EAST PROCESS SAFETY CONFERENCE
Achieve Quick Improvements in Process Safety
Steve Arendt, PE
VP-Global Oil, Gas & Chemicals
+1-281-673-2914
sarendt@absconsulting.com
www.abs-group.com

Achieving Excellence in PSM Through Competency

David A. Moore, PE, CSP, CPSA
President & CEO
And
Cara Hamel
PSM Engineer
AcuTech Group, Inc.
Vienna, Virginia, USA 22182
www.acutech-consulting.com

AcuTech

CCPS MIDDLE EAST PROCESS SAFETY CONFERENCE

October 9-11, 2017
Bahrain International Exhibition and Convention Centre

1

Overview

AcuTech

- What is Process Safety Competency?
- Importance of Strong Competency
  - CCPS Guidelines for Risk Based Process Safety
  - CCPS Vision 20/20
- Key Principles of PSM Competency
- PSM Competency and Culture

2

What is Process Safety Competency?

- Under the RBPS pillar of *Committing to Process Safety*, one of the five elements is "developing, sustaining, and enhancing the organization's process safety competency"

- *CCPS Guidelines for Risk Based Process Safety* states that developing and maintaining process safety competency encompasses three interrelated actions:
  - Continuously improving knowledge and competence
  - Ensuring that the appropriate information is available to those in the organization who need it
  - Consistently applying what has been learned

What is Process Safety Competency?

- Process safety competency is closely related to the "knowledge" and "training" elements of the RBPS system.
- The competence of each individual in an organization is developed and maintained through *training.*
- The competence of an organization depends on the body of *knowledge* increasing, and the application of the body of process knowledge to situations that manage risk and improve plant performance and safety.

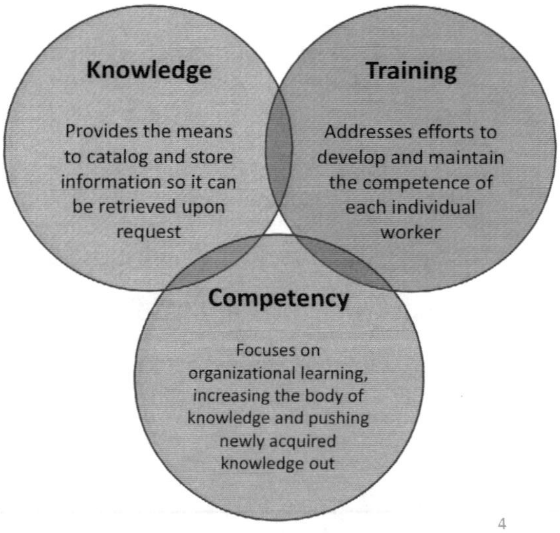

Importance of Strong Competency

AcuTech

CCPS VISION 20/20

- CCPS recognized "Intentional Competency Development" as one of the five tenets of their Vision 20/20
- Competency contributes to Vision 20/20 because:
  - No matter how strong the company culture or management system is, or how well standards are followed, it requires ***competent employees*** to implement those systems and standard
  - ***Intentional competency development is required***, including setting and understanding competency expectations, providing educational resources, and allowing time for people to build competency
  - ***Intentional competency development applies to all levels in the organization***. It starts with leaders visibly leading process safety, and then engineers implementing technical designs and operators knowing their processes and safe operating limits

https://www.aiche.org/ccps/resources/vision-2020/five-industry-tenets/intentional-competency-development

5

Importance of Strong Competency

AcuTech

- Excellence in PSM is not possible to achieve nor sustain without a competent workforce
- "Only competent people can transform information into knowledge" (*CCPS Guidelines for RBPS*)
- Knowledge management, rather than information management, helps organizations understand and manage risk

6

Considerations for PSM Competency

AcuTech

- In order to form a competent workforce and to facilitate the PSM culture, there must be a short and long-term plan in place

7

Key Principles of PSM Competency

AcuTech

Maintain a Dependable Practice	• Establish objectives and set a plan to achieve these objectives • Appoint a champion • Identify corollary benefits • Promote a learning organization
1.Execute activities that help maintain and enhance process safety competency	• Appoint technology stewards • Document knowledge and ensure that information is accessible, and provide structure • Push knowledge to appropriate personnel, update information • Promote person-to-person contact and plan personnel transitions
Evaluate and share results	• Evaluate the utility of existing efforts • Solicit needs from operating units
Adjust plans	• Periodically review the status of process safety competency efforts • Revise the plans as needed to more closely align with the perceived needs

8

PSM Competency and PSM Culture

AcuTech

- "Unleashing" the workforce to achieve best performance under the right PSM culture is a key challenge
- Competency of the workforce has to be an integral part of the process safety management system to be effective and to have a breakthrough to the next level of performance
- This requires the right culture where managers are encouraged to develop a proper mentoring culture and to bring out excellence in their workforce

9

PSM Competency and PSM Culture

AcuTech

- Core Principles of PSM Culture:
  1. Establish an Imperative for Safety
  2. Provide Strong Leadership
  3. Maintain a Sense of Vulnerability
  4. Understand and Act Upon Hazards/Risks
  5. Empower Individuals to Successfully Fulfill their Safety Responsibilities
  6. Defer to Expertise
  7. Ensure Open and Frank Communications
  8. Foster Mutual Trust
  9. Combat the Normalization of Deviance
  10. Learn to Assess and Advance the Culture

10

Summary

- A PSM initiative depends on competency in the workforce
- This must be planned, nurtured, monitored, and improved as required
- Human capital is the most valuable investment.

11

PROCESS IMPROVEMENT INSTITUTE
Risk and reliability specialists

CCPS
MIDDLE EAST
PROCESS SAFETY
CONFERENCE

Four major gaps preventing most companies worldwide from achieving excellent process safety performance

William G. Bridges

PROCESS IMPROVEMENT INSTITUTE
Risk and reliability specialists

CCPS
MIDDLE EAST
PROCESS SAFETY
CONFERENCE

William Bridges

Process Improvement Institute - President

STUDIES

BS & MS Chemical Engineering

Functional safety professional

Process safety expert

Co-invented LOPA

Human factors expert

EXPERIENCE

39 years

13 years in plant, 2 as Operator

+250 Unit PHAs

+8000 PHAs managed

+2000 LOPAs

+1000 SIL Verification/Assessments

+3500 trained PHA/LOPA leaders

+3000 folks trained in GCC area

PUBLICATIONS

Main author

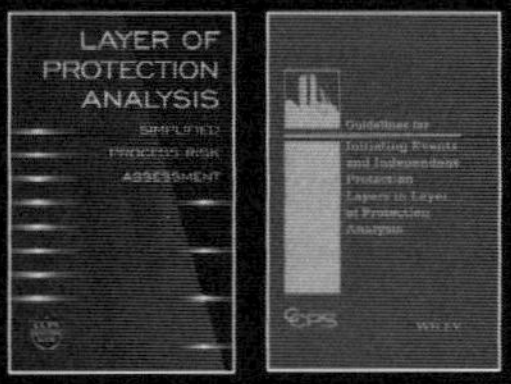

5 CCPS Guidelines book contributing author

+50 Papers presented

PROCESS IMPROVEMENT INSTITUTE
Risk and reliability specialists
CCPS
MIDDLE EAST
PROCESS SAFETY
CONFERENCE
PROCEDURES
Accurate and clear operating and maintenance work instructions
NEAR MISSES
Getting Near Misses reported and investigated
PHA
Of all modes of operation (including startup, shutdown, and online maintenance) and all damage mechanism
HUMAN FACTORS
Addressing human factors missing from most management systems

PROCESS IMPROVEMENT INSTITUTE
Risk and reliability specialists
CCPS
MIDDLE EAST
PROCESS SAFETY
CONFERENCE
PROCEDURES
Accurate and clear operating and maintenance work instructions
NEAR MISSES
Getting Near Misses reported and investigated
PHA
Of all modes of operation (including startup, shutdown, and online maintenance)
HUMAN FACTORS
Addressing human factors missing from most management systems

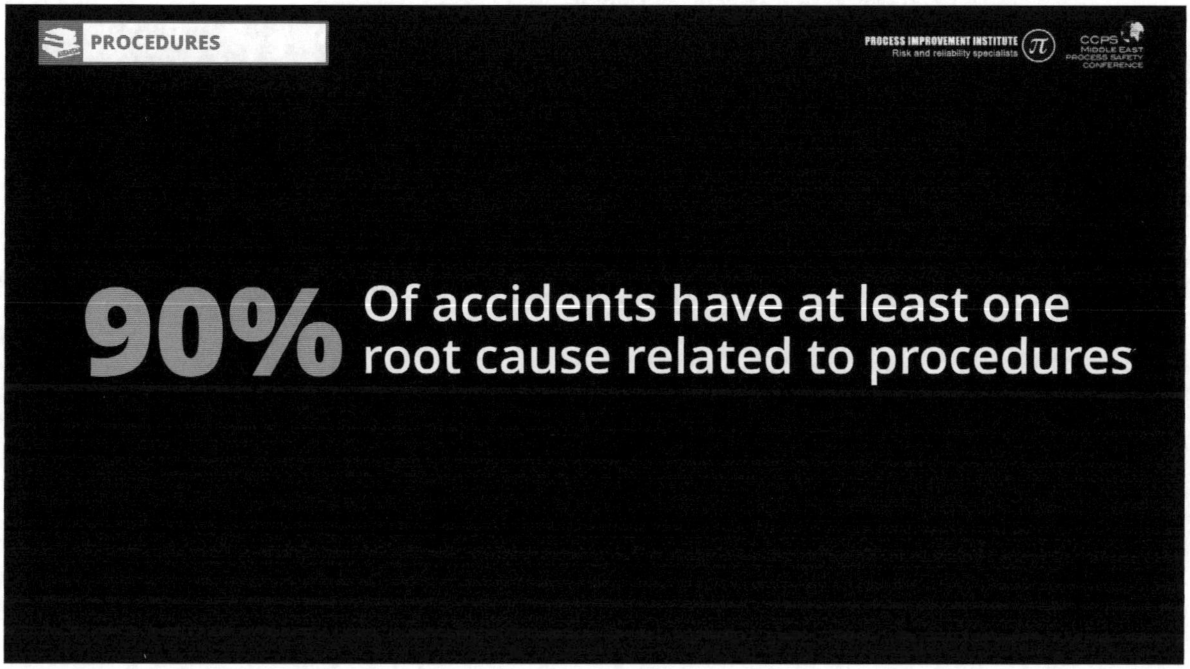
PROCEDURES
PROCESS IMPROVEMENT INSTITUTE
Risk and reliability specialists
CCPS
MIDDLE EAST
PROCESS SAFETY
CONFERENCE
90%
Of accidents have at least one root cause related to procedures

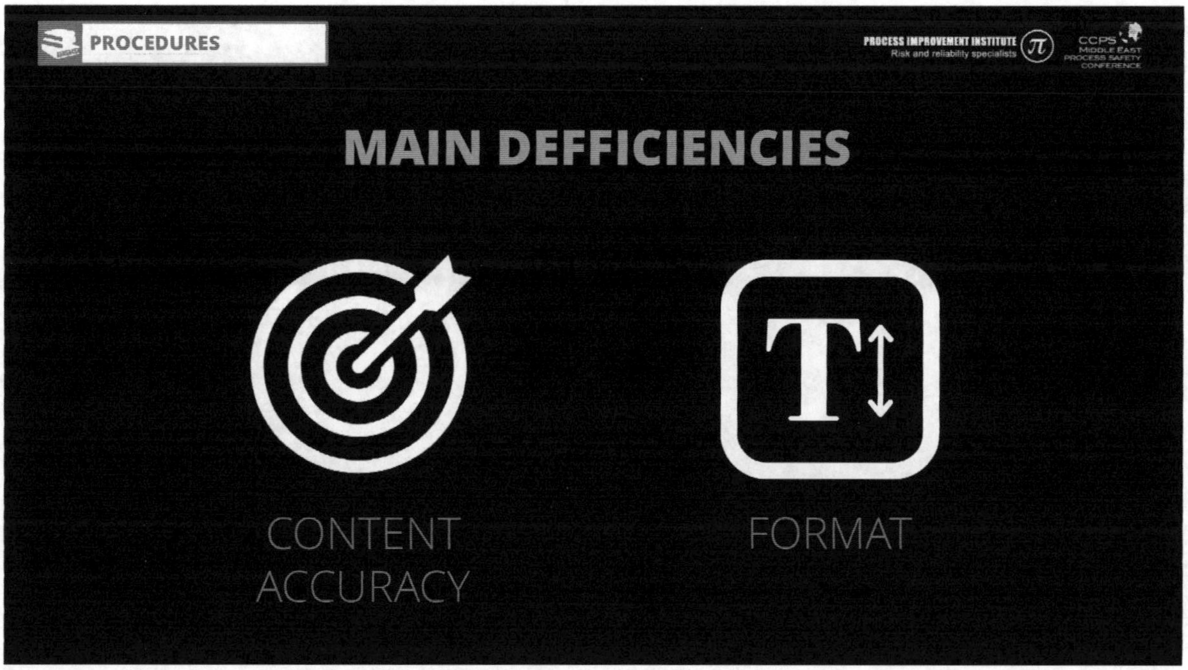
PROCEDURES
PROCESS IMPROVEMENT INSTITUTE
Risk and reliability specialists
CCPS
MIDDLE EAST
PROCESS SAFETY
CONFERENCE
MAIN DEFFICIENCIES
CONTENT ACCURACY
FORMAT

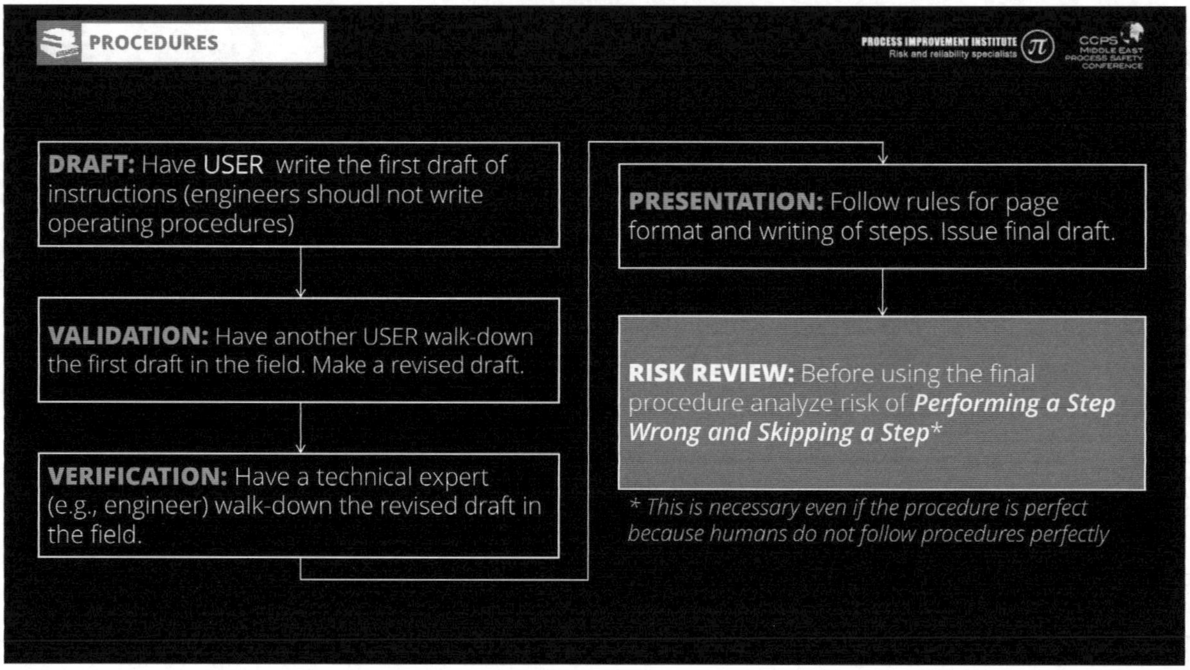
PROCEDURES
PROCESS IMPROVEMENT INSTITUTE
Risk and reliability specialists
CCPS
MIDDLE EAST
PROCESS SAFETY
CONFERENCE
DRAFT: Have USER write the first draft of instructions (engineers shoudl not write operating procedures)
VALIDATION: Have another USER walk-down the first draft in the field. Make a revised draft.
VERIFICATION: Have a technical expert (e.g., engineer) walk-down the revised draft in the field.
PRESENTATION: Follow rules for page format and writing of steps. Issue final draft.
RISK REVIEW: Before using the final procedure analyze risk of Performing a Step Wrong and Skipping a Step*
* This is necessary even if the procedure is perfect because humans do not follow procedures perfectly

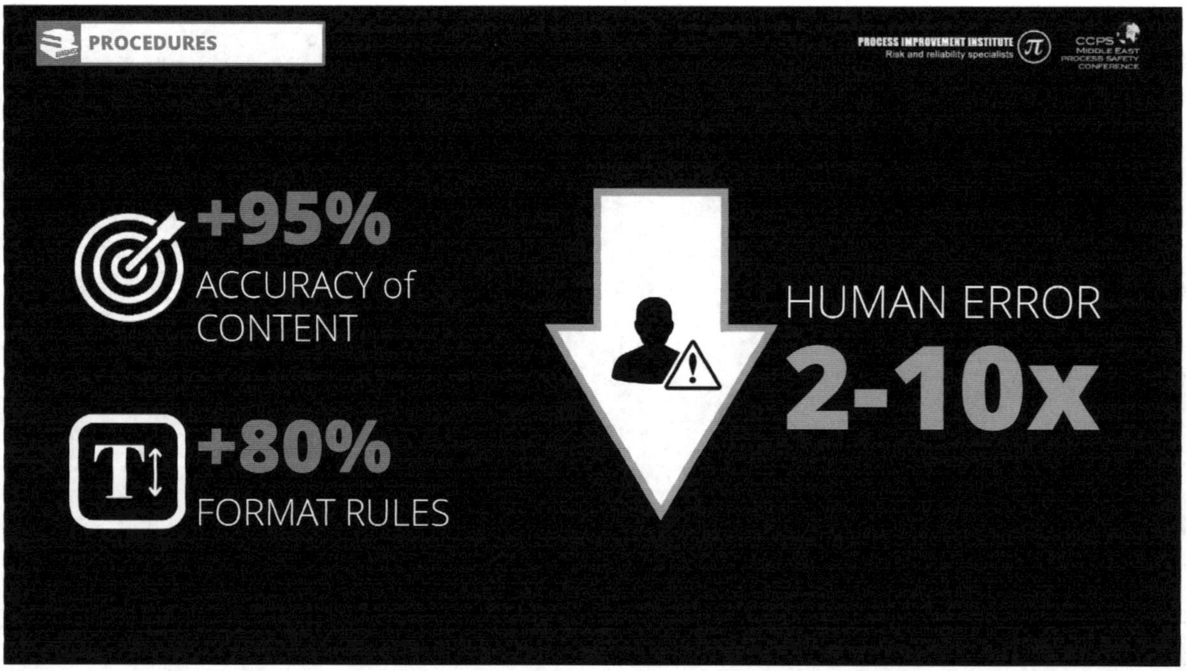
PROCEDURES
PROCESS IMPROVEMENT INSTITUTE
Risk and reliability specialists
CCPS
MIDDLE EAST
PROCESS SAFETY
CONFERENCE
+95%
ACCURACY of CONTENT
+80%
FORMAT RULES
HUMAN ERROR
2-10x

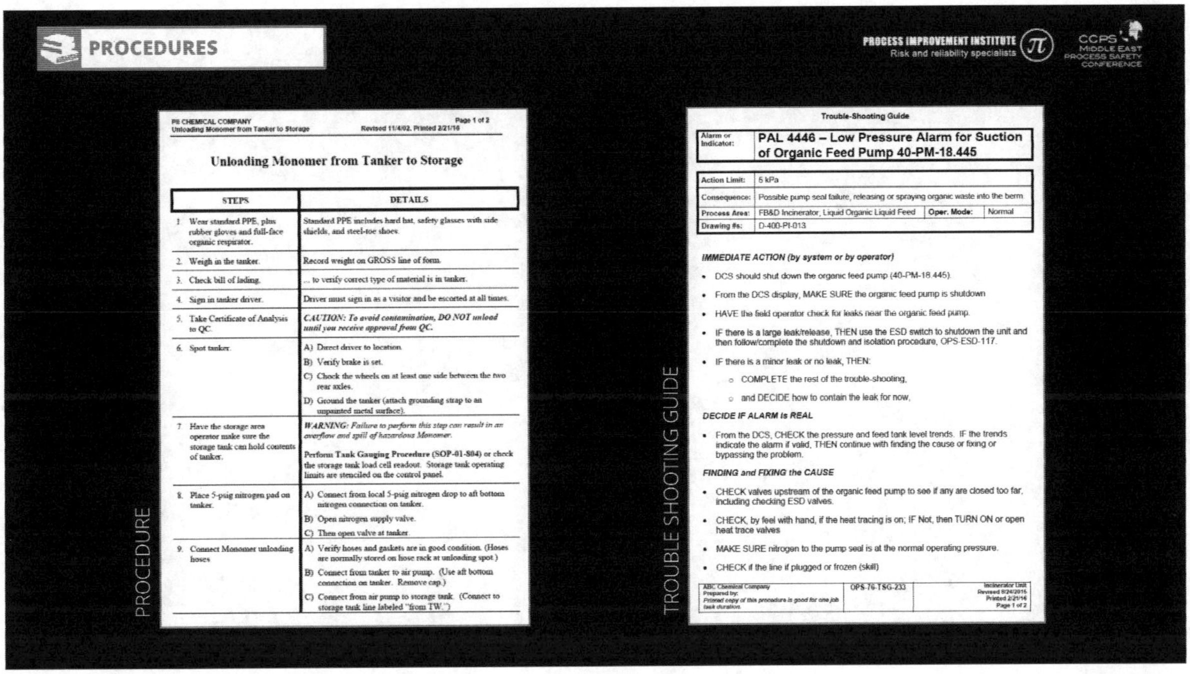
PROCEDURES
PROCESS IMPROVEMENT INSTITUTE
Risk and reliability specialists
CCPS
MIDDLE EAST
PROCESS SAFETY
CONFERENCE
PROCEDURE
PE CHEMICAL COMPANY
Unloading Monomer from Tanker to Storage
Page 1 of 2
Revised 11/4/02, Printed 2/21/16
Unloading Monomer from Tanker to Storage
STEPS
DETAILS
1. Wear standard PPE, plus rubber gloves and full-face organic respirator.
Standard PPE includes hard hat, safety glasses with side shields, and steel-toe shoes.
2. Weigh in the tanker.
Record weight on GROSS line of form.
3. Check bill of lading.
... to verify correct type of material is in tanker.
4. Sign in tanker driver.
Driver must sign in as a visitor and be escorted at all times.
5. Take Certificate of Analysis to QC.
CAUTION: To avoid contamination, DO NOT unload until you receive approval from QC.
6. Spot tanker.
A) Direct driver to location.
B) Verify brake is set.
C) Chock the wheels on at least one side between the two rear axles.
D) Ground the tanker (attach grounding strap to an unpainted metal surface).
7. Have the storage area operator make sure the storage tank can hold contents of tanker.
WARNING: Failure to perform this step can result in an overflow and spill of hazardous Monomer.
Perform Tank Gauging Procedure (SOP-01-804) or check the storage tank load cell readout. Storage tank operating limits are stenciled on the control panel.
8. Place 5-psig nitrogen pad on tanker.
A) Connect from local 5-psig nitrogen drop to aft bottom nitrogen connection on tanker.
B) Open nitrogen supply valve.
C) Then open valve at tanker.
9. Connect Monomer unloading hoses
A) Verify hoses and gaskets are in good condition. (Hoses are normally stored on hose rack at unloading spot.)
B) Connect from tanker to air pump. (Use aft bottom connection on tanker. Remove cap.)
C) Connect from air pump to storage tank. (Connect to storage tank line labeled "from TW.")
TROUBLE SHOOTING GUIDE
Trouble-Shooting Guide
Alarm or Indicator:
PAL 4446 – Low Pressure Alarm for Suction of Organic Feed Pump 40-PM-18.445
Action Limit: 5 kPa
Consequence: Possible pump seal failure, releasing or spraying organic waste into the berm
Process Area: FB&D Incinerator, Liquid Organic Liquid Feed
Oper. Mode: Normal
Drawing #s: D-400-PI-013
IMMEDIATE ACTION (by system or by operator)
• DCS should shut down the organic feed pump (40-PM-18.445).
• From the DCS display, MAKE SURE the organic feed pump is shutdown
• HAVE the field operator check for leaks near the organic feed pump.
• IF there is a large leak/release, THEN use the ESD switch to shutdown the unit and then follow/complete the shutdown and isolation procedure, OPS-ESD-117.
• IF there is a minor leak or no leak, THEN:
○ COMPLETE the rest of the trouble-shooting,
○ and DECIDE how to contain the leak for now,
DECIDE IF ALARM Is REAL
• From the DCS, CHECK the pressure and feed tank level trends. IF the trends indicate the alarm if valid, THEN continue with finding the cause or fixing or bypassing the problem.
FINDING and FIXING the CAUSE
• CHECK valves upstream of the organic feed pump to see if any are closed too far, including checking ESD valves.
• CHECK, by feel with hand, if the heat tracing is on; IF Not, then TURN ON or open heat trace valves
• MAKE SURE nitrogen to the pump seal is at the normal operating pressure.
• CHECK if the line if plugged or frozen (skill)
ABC Chemical Company
Prepared by:
Printed copy of this procedure is good for one job task duration.
OPS-76-TSG-233
Incinerator Unit
Revised 8/24/2015
Printed 2/21/16
Page 1 of 2

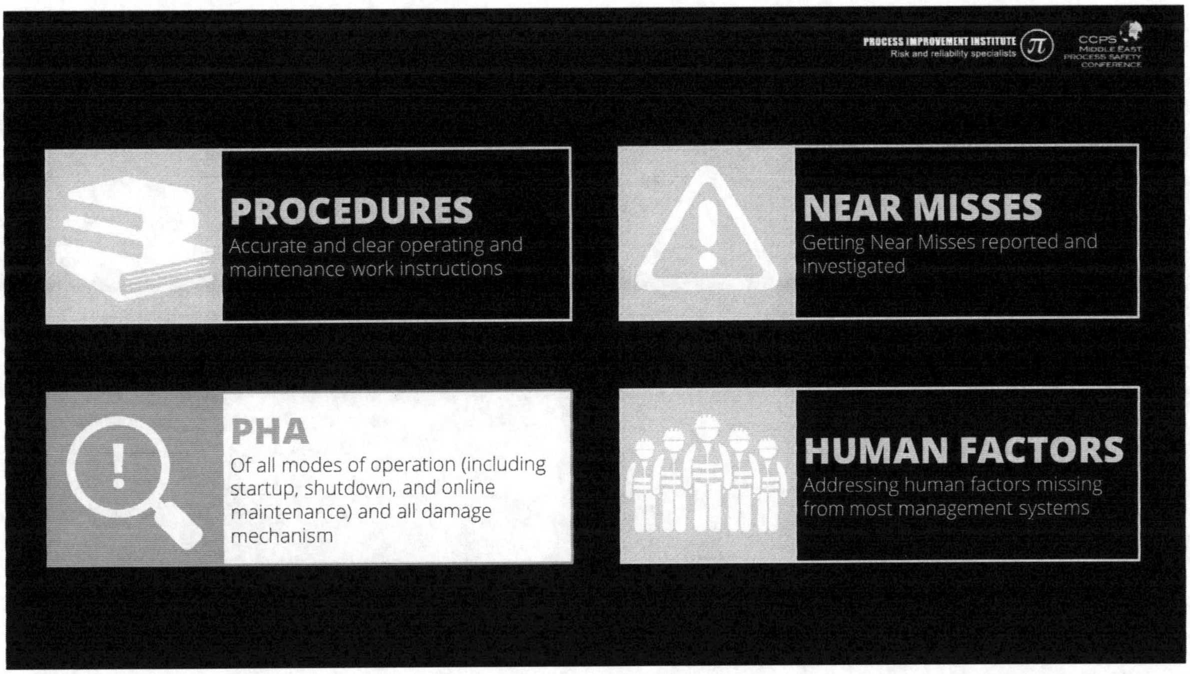
PROCESS IMPROVEMENT INSTITUTE
Risk and reliability specialists
CCPS
MIDDLE EAST
PROCESS SAFETY
CONFERENCE
PROCEDURES
Accurate and clear operating and maintenance work instructions
NEAR MISSES
Getting Near Misses reported and investigated
PHA
Of all modes of operation (including startup, shutdown, and online maintenance) and all damage mechanism
HUMAN FACTORS
Addressing human factors missing from most management systems

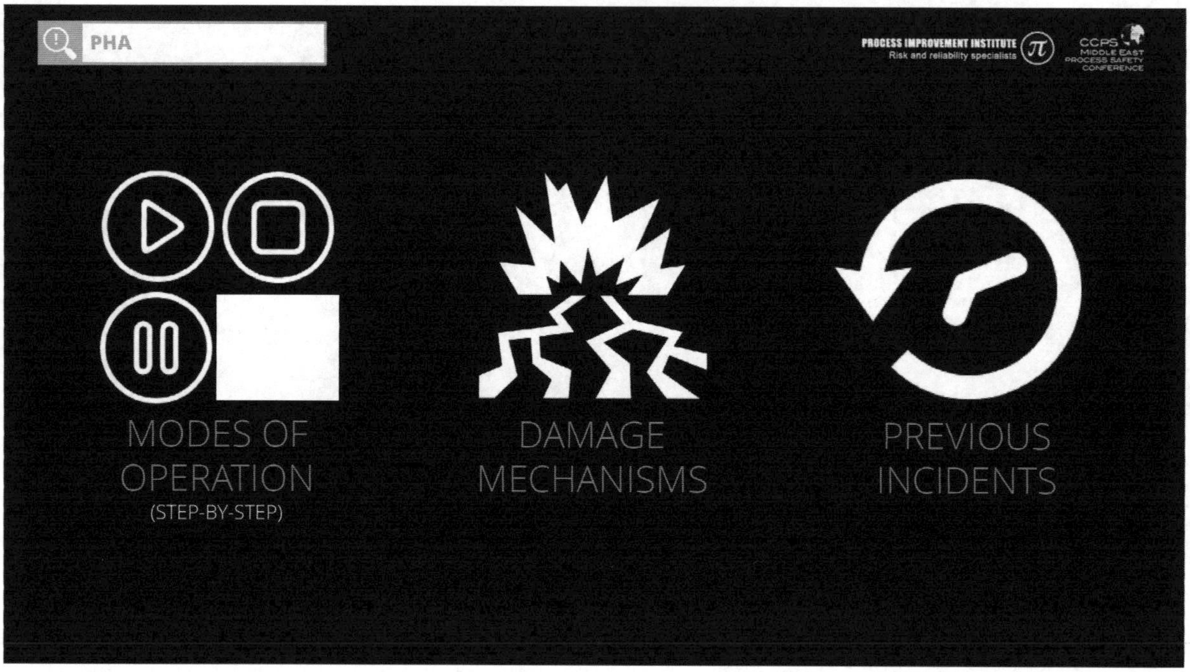
PHA
PROCESS IMPROVEMENT INSTITUTE
Risk and reliability specialists
CCPS MIDDLE EAST PROCESS SAFETY CONFERENCE
MODES OF OPERATION
(STEP-BY-STEP)
DAMAGE MECHANISMS
PREVIOUS INCIDENTS

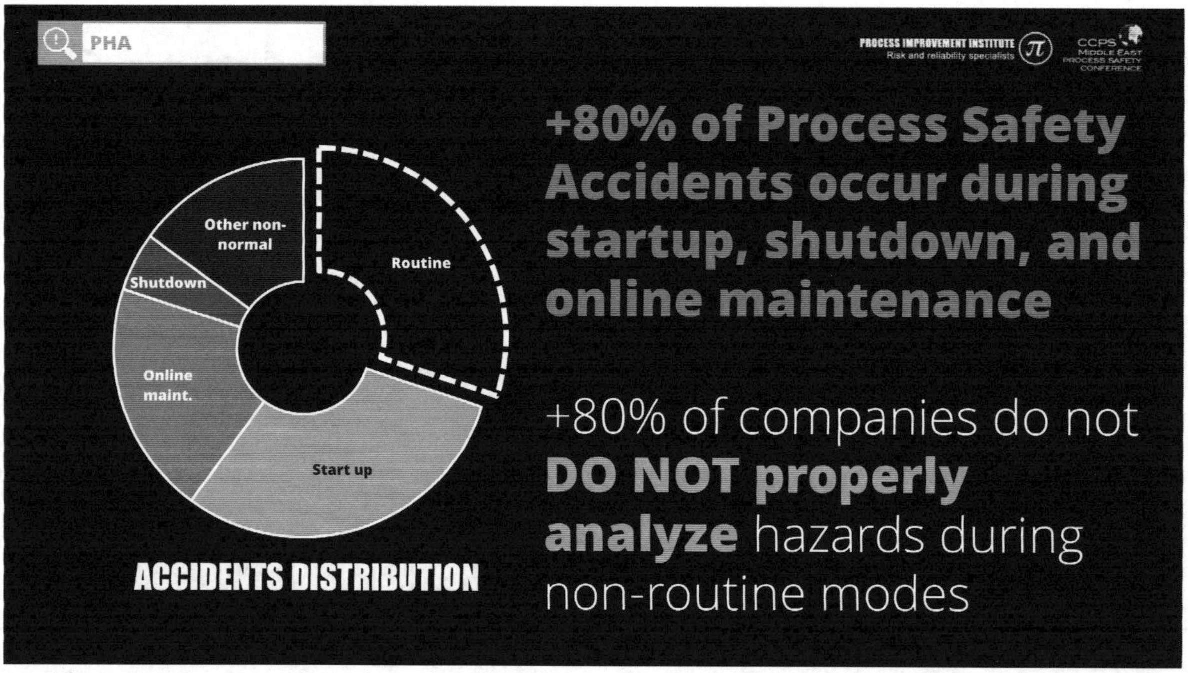
PHA
PROCESS IMPROVEMENT INSTITUTE
Risk and reliability specialists
CCPS MIDDLE EAST PROCESS SAFETY CONFERENCE
Other non-normal
Shutdown
Routine
Online maint.
Start up
ACCIDENTS DISTRIBUTION
+80% of Process Safety Accidents occur during startup, shutdown, and online maintenance
+80% of companies do not DO NOT properly analyze hazards during non-routine modes

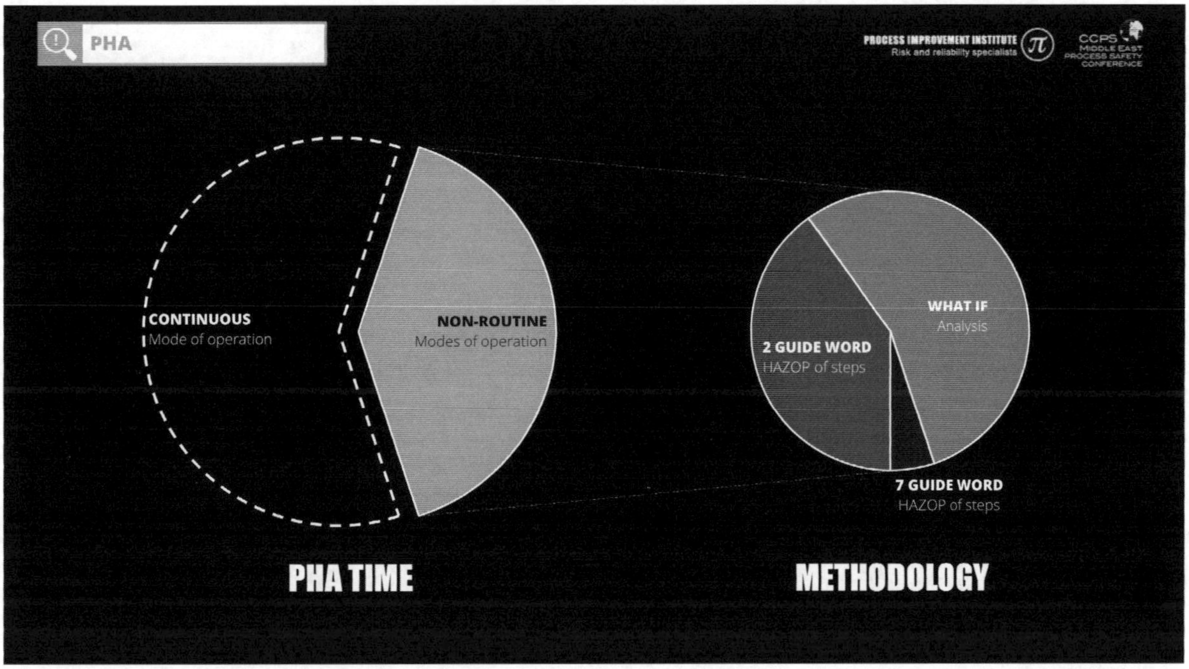
PHA
PROCESS IMPROVEMENT INSTITUTE
Risk and reliability specialists
CCPS
MIDDLE EAST
PROCESS SAFETY
CONFERENCE
CONTINUOUS
Mode of operation
NON-ROUTINE
Modes of operation
WHAT IF
Analysis
2 GUIDE WORD
HAZOP of steps
7 GUIDE WORD
HAZOP of steps
PHA TIME
METHODOLOGY

PHA
PROCESS IMPROVEMENT INSTITUTE
Risk and reliability specialists
CCPS
MIDDLE EAST
PROCESS SAFETY
CONFERENCE
70%
PHAs do not cover Damage Mechanisms in each node*
*PHAs audited by PII

Example A – Referencing Generic Tables for Typical Causes and Typical Safeguards

Dev#	Deviation	Causes	Consequences	Existing Safeguards	Recommendations
9.10	Loss of containment	Accelerate corrosion External fire High pressure (linked from 9.7) Typical causes of loss of containment (see Table A.1)	Release to atmosphere leading to potential injury of workers and/or community	Pressure Relief Valve's 231A, B, C on Reactor and no valves in line to reactor Most lines / connections are welded construction; only a few flanges Drills conducted each year on evacuation, rescue and isolation Emergency Response personnel are trained at SABIC FTC Generic safeguards protecting against or mitigating process material releases (see Table A.2)	Safety 7. Consider changing the ITPM schedule for managing most PSVs whose inspections are too infrequent based on industry standards and best practice. For instance, the current inspection frequency for PSV-8220 on the Ammonia Receiver (I-2005G) is 9 years, whereas, consensus codes typically recommend testing/inspection every 1-4 years for PSVs in highly toxic services.

Example B – Specific Listing of Causes and Safeguards

1.9	Loss of containment	Corrosion/erosion External fire and/or flame impingement Gasket, packing, or seal failure Improper maintenance Material defect Operator failing to close or inadvertently opening a valve to the atmosphere (e.g., a valve at a hose connection) Railcar inadvertently derailed Valve leaking to the atmosphere High pressure (linked from 1.5) Acid corrosion caused by high concentration of water (linked from 1.8) High ambient temperature External impact (such as from a mini-engine or another railcar)	Catastrophic release of chlorine from a ruptured railcar Steady release of chlorine from a ruptured connection Steady release of chlorine from a leaky connection High pressure caused by thermal expansion of liquid chlorine if railcar is also over-full	Chlorine repair kit Derailer and warning flag to prevent impact by a mini-engine or another railcar Limited vehicular access to area Maintenance/operator response as required, including isolation if needed Operator periodically monitoring the railcar valves while unloading Personal protective equipment in the area Plugs installed in all chlorine valves to the atmosphere when the valves are not in use Relief valve on each railcar for mitigating releases caused by overpressure Supplier maintenance of railcars (per strictly enforced US DOT requirements) Video monitoring of the unloading area Concrete crossties on rail spur Dike preventing any combustibles spilled nearby from reaching the unloading rack area Concrete railroad ties in chlorine unloading area to prevent fires near railcar	10. Consider installing a chlorine detection system in the unloading and vaporizing area to help detect chlorine releases (especially at likely release points) 11. Verify that periodic maintenance and inspections are being performed in accordance with Chlorine Institute recommendations 12. Review the drainage system for the unloading are, and identify the areas that may be affected by a large chlorine release 13. Consider prohibiting the use of heavy equipment (e.g., cranes) in the unloading and vaporizing area unless special precautions to prevent equipment damage are enacted 33. Consider providing a high pressure alarm for each vaporizer 40. Consider providing a water deluge system in the unloading area to help mitigate chlorine releases from the railcar

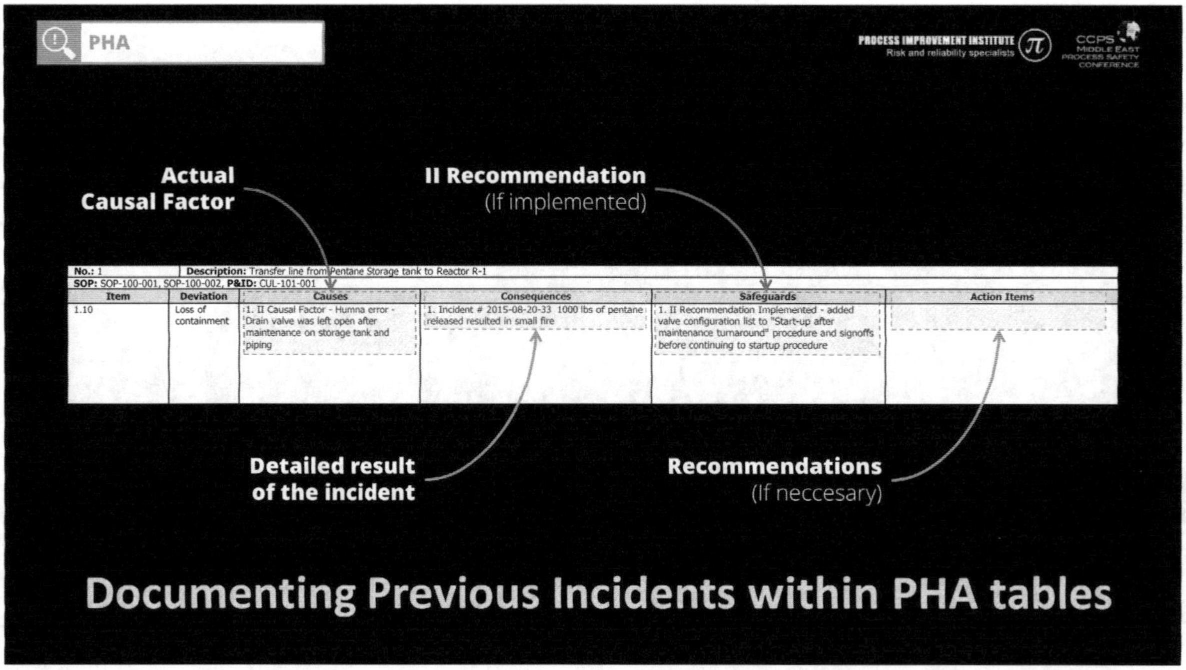

PROCESS IMPROVEMENT INSTITUTE
Risk and reliability specialists
CCPS
MIDDLE EAST
PROCESS SAFETY
CONFERENCE
PROCEDURES
Accurate and clear operating and maintenance work instructions
NEAR MISSES
Getting Near Misses reported and investigated
PHA
Of all modes of operation (including startup, shutdown, and online maintenance)
HUMAN FACTORS
Addressing human factors missing from most management systems

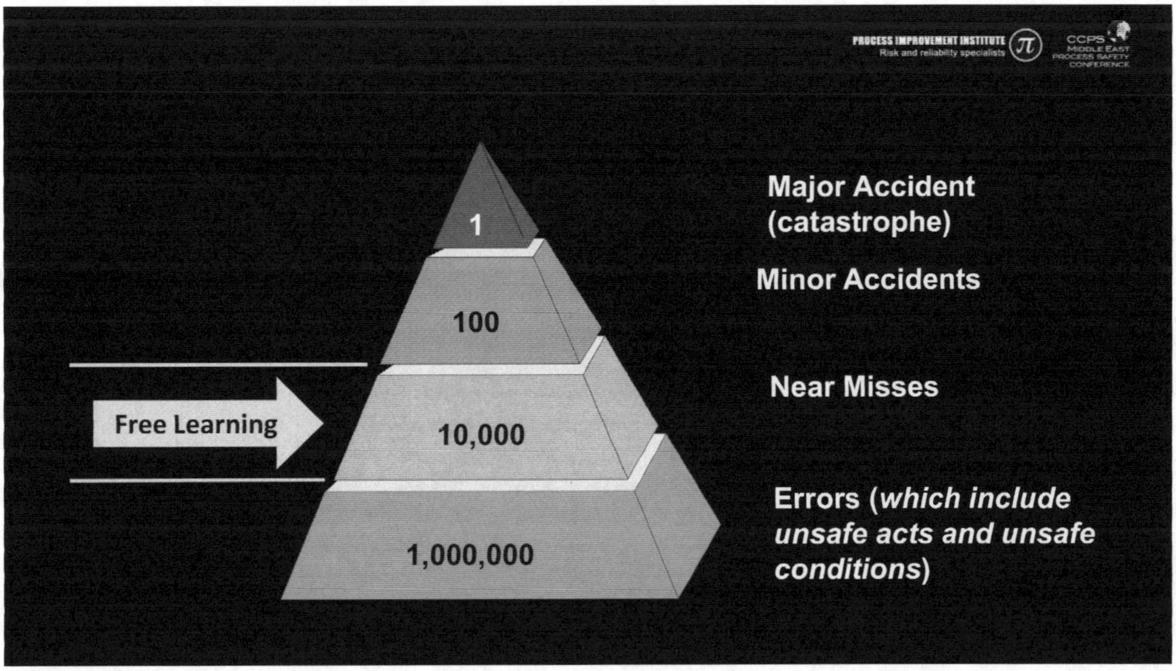
PROCESS IMPROVEMENT INSTITUTE
Risk and reliability specialists
CCPS
MIDDLE EAST
PROCESS SAFETY
CONFERENCE
1
100
10,000
1,000,000
Free Learning
Major Accident (catastrophe)
Minor Accidents
Near Misses
Errors (*which include unsafe acts and unsafe conditions*)

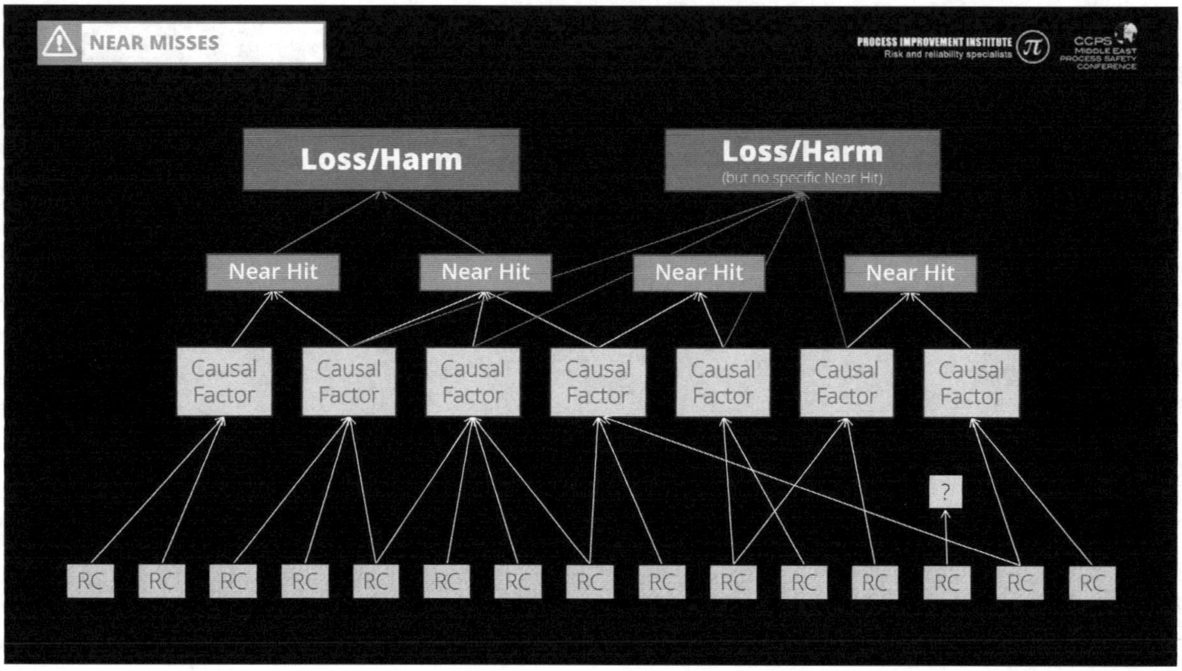
NEAR MISSES
PROCESS IMPROVEMENT INSTITUTE
Risk and reliability specialists
CCPS
MIDDLE EAST
PROCESS SAFETY
CONFERENCE
Loss/Harm
Loss/Harm
(but no specific Near Hit)
Near Hit
Near Hit
Near Hit
Near Hit
Causal Factor
Causal Factor
Causal Factor
Causal Factor
Causal Factor
Causal Factor
Causal Factor
?
RC RC RC RC RC RC RC RC RC RC RC RC RC RC RC

NEAR MISSES
PROCESS IMPROVEMENT INSTITUTE
Risk and reliability specialists
CCPS
MIDDLE EAST
PROCESS SAFETY
CONFERENCE
15-20 = NEAR MISS INVESTIGATION / ACCIDENT
+90% Chance of finding problems and correcting them

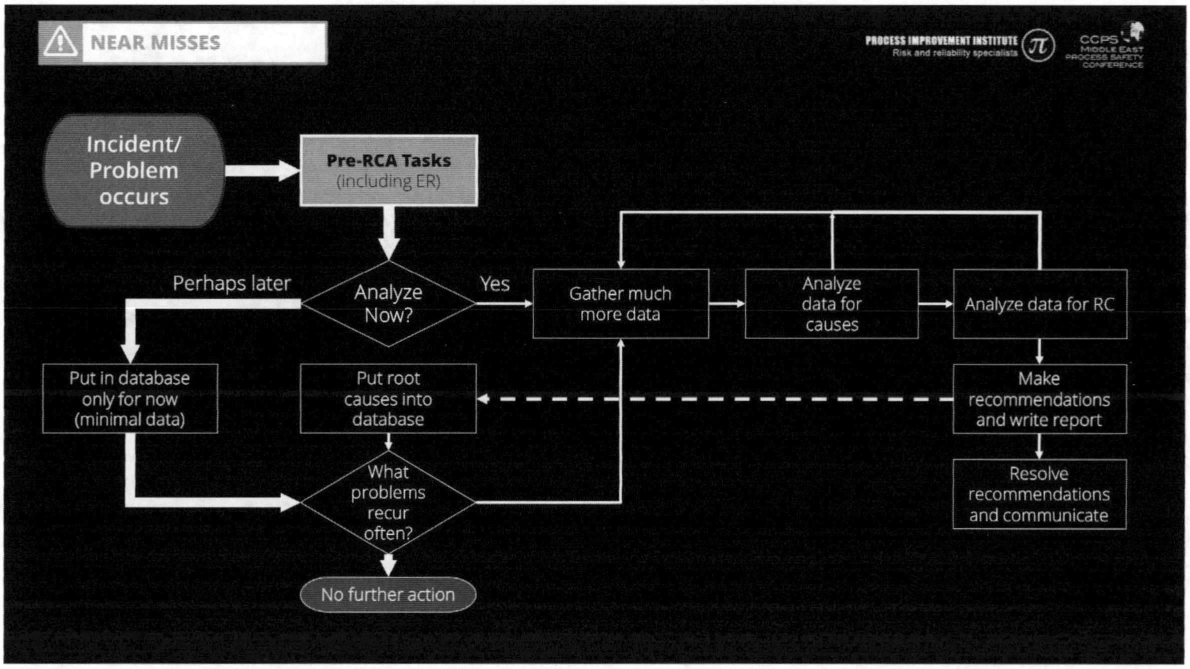
NEAR MISSES
PROCESS IMPROVEMENT INSTITUTE
Risk and reliability specialists
CCPS
MIDDLE EAST
PROCESS SAFETY
CONFERENCE
Incident/ Problem occurs
Pre-RCA Tasks (including ER)
Analyze Now?
Perhaps later
Yes
Gather much more data
Analyze data for causes
Analyze data for RC
Put in database only for now (minimal data)
Put root causes into database
Make recommendations and write report
Resolve recommendations and communicate
What problems recur often?
No further action

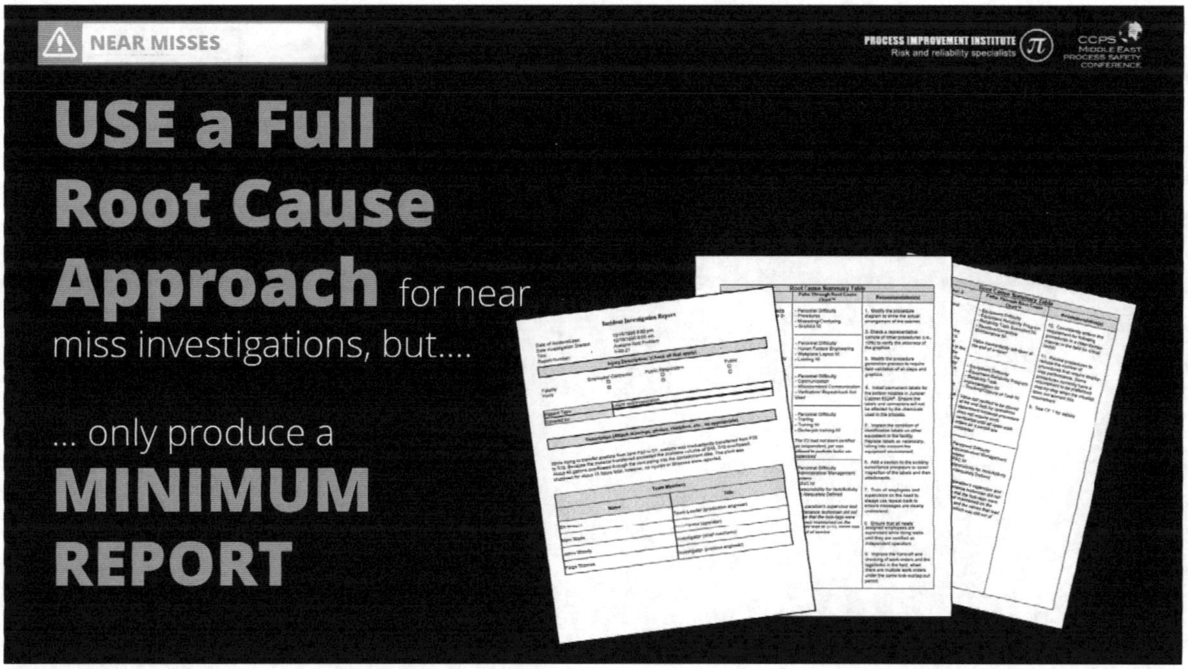
NEAR MISSES
PROCESS IMPROVEMENT INSTITUTE
Risk and reliability specialists
CCPS
MIDDLE EAST
PROCESS SAFETY
CONFERENCE
USE a Full Root Cause Approach for near miss investigations, but....
... only produce a MINIMUM REPORT
Incident Investigation Report
Root Cause Summary Table
Root Cause Summary Table

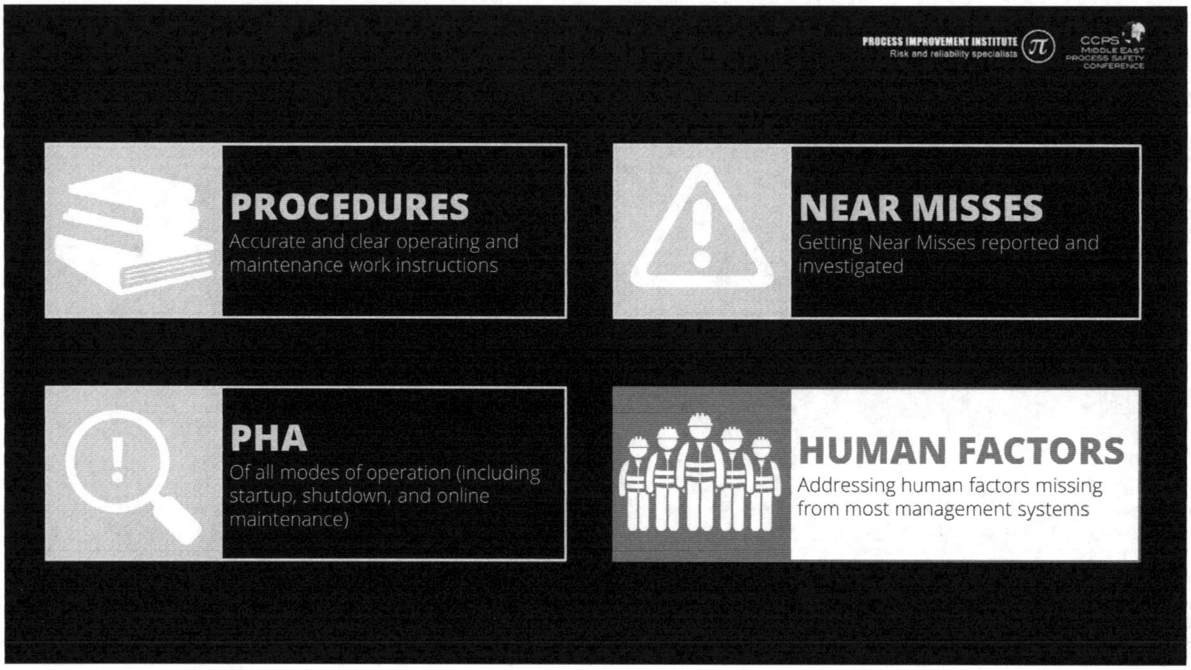
PROCESS IMPROVEMENT INSTITUTE
Risk and reliability specialists
CCPS
MIDDLE EAST
PROCESS SAFETY
CONFERENCE
PROCEDURES
Accurate and clear operating and maintenance work instructions
NEAR MISSES
Getting Near Misses reported and investigated
PHA
Of all modes of operation (including startup, shutdown, and online maintenance)
HUMAN FACTORS
Addressing human factors missing from most management systems

HUMAN FACTORS
PROCESS IMPROVEMENT INSTITUTE
Risk and reliability specialists
CCPS
MIDDLE EAST
PROCESS SAFETY
CONFERENCE
99%
of accidental losses begin with a human error*
(except for natural disasters)
*supported by data from more than 1500 investigations

HUMAN FACTORS

PROCESS IMPROVEMENT INSTITUTE
Risk and reliability specialists

CCPS Middle East Process Safety Conference

Root causes of accidents are **management system weaknesses**

(Center for Chemical Process Safety, American Institute for Chemical Engineers, "Guidelines for Investigating Chemical Process Incidents", 2003) – US OSHA agrees

WEAK MANAGEMENT SYSTEMS → **HUMAN ERROR** → **Near Misses** → **ACCIDENTS**

HUMAN FACTORS

PROCESS IMPROVEMENT INSTITUTE
Risk and reliability specialists

CCPS Middle East Process Safety Conference

HUMAN FACTORS CATEGORIES

Human factor category	Human factor issue/level	Multiplier for cognitive & diagnosis errors
Available time (Includes staffing issues) For responses only	Inadequate time	P(failure) = 100%
	Barely adequate time (~2/3 × nominal)	10
	Nominal time	1
	Extra time (Between 1 and 2 and > 20min)	0.1
	Expansive time (>2x nominal and > 20min)	0.01
Stress/Stressor (Includes staffing issues)	Extreme	5
	High	2
	Nominal	1
Complexity and task design	Highly complex	5
	Moderately complex	2
	Nominal	1
	Obvious diagnosis	0.1
Experience/ Training	Low	10
	Nominal	1
	High	0.5
Procedures **1**	Not available	20
	Incomplete	10
	Available, but poor	5
	Nominal	0.5
	Diagnostic/Symptom oriented	1

Human factor category	Human factor issue/level	Multiplier for cognitive & diagnosis errors
Human-Machine interface (includes tools)	Missing/Misleading	20
	Poor	10
	Nominal	1
	Good	0.5
Fitness for duty **3**	Unfit (High fatigue level, ilness, strong medication, not physically capable of job today)	20
	Degraded fitness	5
	Nominal	1
Work processes and supervision	Poor	2
	Nominal	1
	Good	0.8
Work environment	Extreme	5
	Good	1
Communication **2**	No communication or system interference/damage	10
	No standard for verbal communication rules	5
	Well implemented and practiced standard	1

HUMAN FACTORS

PROCESS IMPROVEMENT INSTITUTE
Risk and reliability specialists

CCPS MIDDLE EAST PROCESS SAFETY CONFERENCE

HUMAN FACTORS TYPICALLY MISSING FROM PROCESS SAFETY

1. **Best practices for content and format of OP procedures**
2. **Verbal Communication Standard** (repeat back, etc.)
3. **Fitness for duty** (fatigue management, etc.)
4. **Task design to match human** (includes work environment)
5. **Human-System Interface** (displays, labels, handheld prompts, etc.)
6. **Staffing considerations for error reduction**

PROCESS IMPROVEMENT INSTITUTE
Risk and reliability specialists

CCPS MIDDLE EAST PROCESS SAFETY CONFERENCE

Closing gaps and finding missing scenarios has
greater than 100:1 payback

saudi aramco

entrepreneurship

Evergreening - Process Safety Management of Flare and Relief Systems

CCPS/MEPEC 2017

Umar Abbasi - P&CSD Saudi Aramco

where energy is opportunity™

Accidents - Flare and Relief Systems - Milford Haven

Milford Haven FCCU flare system – 1994

- The immediate cause of the explosion was the failure of the flare system due to flammable hydrocarbon being continuously pumped into the debutanizer that had its outlet control valve closed
- One of the underlying cause was the modification to the flare knock-out pump out system carried out without assessing all the consequences

Accidents - Flare and Relief Systems - Texas City

Texas City Explosion - 2005

Blowdown system was an unsafe design; it was originally installed in the 1950s, and had never been connected to a flare system to safely contain liquids and combust flammable vapors released from the process.

3

Saudi Aramco: Confidential

أرامكو السعودية saudi aramco

Evergreening (Objectives)

- Evergreening Program at Saudi Aramco facilities is to efficiently manage the flare and relief systems to prevent unsafe operations endangering the life of operating personnel

- The Evergreening framework is a data and change management system which ensures that flare and relief data remains complete, accurate and up-to-date

4

Saudi Aramco: Confidential

أرامكو السعودية saudi aramco

Evergreening (Scope)

Following elements form the pillars for the Evergreening framework;

Note: For the Evergreening program, the focus areas (elements) are interrelated and dependent upon one another, and failure to properly manage any one of these elements compromises the integrity of the entire evergreening program.

Evergreening (Elements)

Process Safety Information

Process Safety Information requires complete compilation of information related to plant flare and relief systems;

- PSV/flare datasheets
- Relief/flare studies
- Softwares used for flare and relief system documentation/analysis
- PFD/P&IDs
- Equipment & instrument datasheets, access to i-plant/doc libraries etc..........

Evergreening (Elements)

Compliance with Standards

External and internal standards compliance is necessary. Incidents in the past could have been avoided by compliance with API 521 with regards to blow down drum sizing.

Evergreening (Elements)

Training

The following key principles should be addressed when developing, evaluating, or improving any management system for the training element:

- Identify what training is needed
- Provide effective training

Evergreening (Elements)

PHA/Relief System Analysis

- Design basis for all pressure relief devices, depressuing valves, flare headers and stacks, flare knockout drums and pumps and all elements of a flare system should be documented so that it can be ensured that the plant is adequately protected
- PHA actions should be tracked
- Any changes in plant which can have an impact on the flare and relief systems e.g. capacity should flag the need for re-evaluating the flare and relief systems adequacy

Relief Systems Analysis Workflow

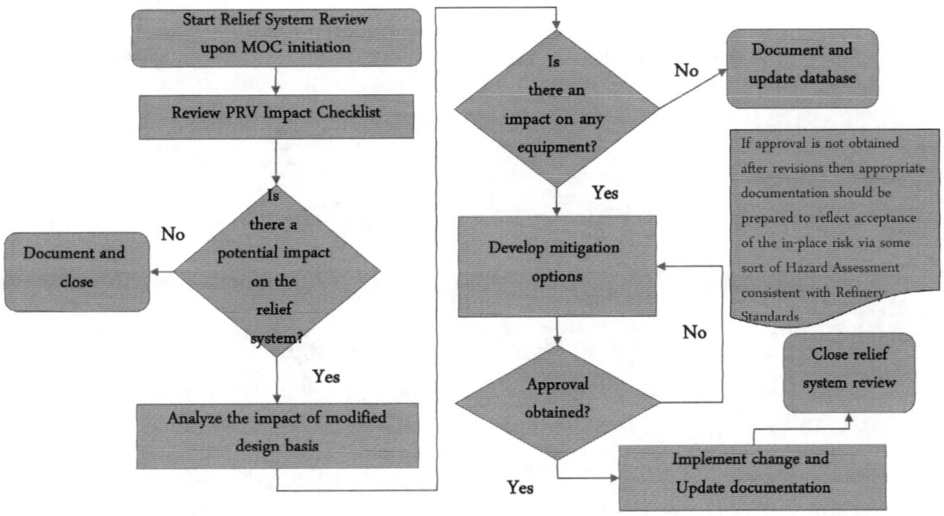

Excellence in Pressure Relief Systems Management_Marshall et al_2011

Evergreening (Elements)

Management of Change (MOC)

The employer shall establish and implement written procedures to manage changes (except for "replacements in kind") to the flare and relief system. The procedures shall assure that the following considerations are addressed prior to any change:

- Impact of change on safety and health of flare and relief systems
- Modifications to operating procedures
- Necessary time period for the change
- Authorization requirements for the proposed change

Evergreening (Elements)

Integrity

This includes;

- Codes and standards
- Equipment maintenance
- Inspection and testing, controlling and managing deficiencies and so on.

Note: It is not just maintenance, although maintenance is a major part of an integrity program. Information is required to identify the code or standard for the design and construction of the flare and relief systems

Evergreening (Elements)

Review/ Audit

- An audit protocol will be prepared to evaluate the performance of the Evergreening systems.
- Audits will be conducted throughout the development and implementation of the Evergreening system.
- Upon completion, a report must be distributed to appropriate parties for follow-up action.

Risk Management (Bow-Tie Approach)

The Bow - Tie approach will be used to assess the effectiveness of Evergreening Elements for any potential accident scenario using the typical risk management control effectiveness rating:

Control Effectiveness Assessment

	Rating	Description
Overall Control Strength	Effective	- Evidence of existence of key controls - Key controls address the related risks - The key controls provide reasonable assurance that the risk will not occur, or if it does occur the control will help detect it timely and minimise the impact - The majority of key controls (>75%) are effective - No key control is rated as 'needs improvement' or 'ineffective'
	Needs Improvement	- The control issues raised have been acknowledged and have appropriate action plans in place - Key controls provide some assurance that the risk will not occur, or if it does it will be detected timely to minimise the impact - The key controls are generally detective rather than preventative in nature
	Ineffective	- Significant issues have been identified in the control environment and require immediate management attention - Key controls provide insufficient/little assurance that the risk event will not occur, or if it does that it will be detected timely to minimise the impact

Accidents - Flare and Relief Systems - Milford Haven

1. Milford Haven Incident

2.Source

Overfilling of Flare KO Drum led to liquid flow into the 30" flare line

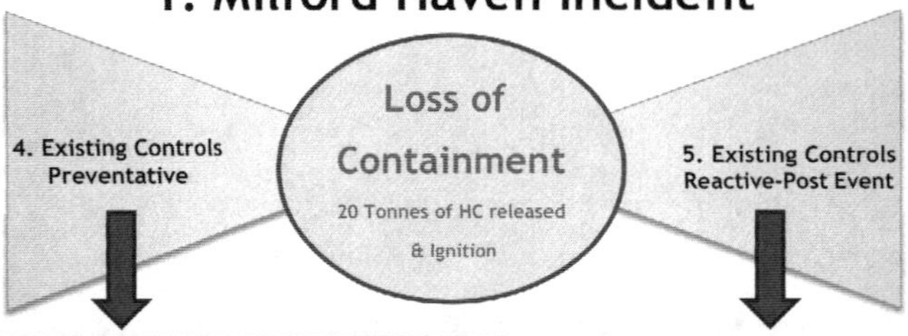

3. Underlying Cause

Large quantities of HC from compressor KO drums by temporarily connecting hoses b/w process vessels and flare header. This added more liquid into the flare KO drum

Failure of Preventive Controls	Evergreening Element
Flare header not designed for liquid weight	PHA Relief Systems Analysis
Alarm prioritisation not very efficient Alarm flooding added to confusion of operations	Integrity
Faulty valve at bottom of the debutaniser - Known corrosion in flare line	Integrity
No MOC carried out to use 2 flexible hoses to drain compressor knock out drums	MOC
No decision to shutdown despite multiple trips	Training
Failure to step back to take a wider view	Training

Failure of Reactive Controls
Door of blast proof control room left open
Lack of fire water for long duration scenarios
Multiple simultaneous fires in 3 refineries

Accidents - Flare and Relief Systems - Texas City Explosion

2. Texas City Explosion

2.Source

Overflowing Hydrocarbon Inventory from the Raffinate Splitter

3. Underlying Cause

Inadequate HAZID skills
Lack of underpinning knowledge
Failure to follow procedures

Failure of Preventive Controls	Evergreening Element
Flare not used	Compliance with Standards/PHA Relief Systems Analysis
No up to date relief study - design basis unclear Capacity of blowdown drum was exceeded	PHA/ Relief Systems Analysis
Inadequate HAZID skills Lack of underpinning knowledge Failure to follow procedures	Compliance with Standards/PHA Relief Systems Analysis
Pre-start-up review not performed Procedural compliance not checked Supervisor offsite	Compliance with Standards/PHA MOC
Blowdown drum modified without rigorous MOC	MOC
No effective handover between shifts Unit alarm not sounded	MOC

Failure of Reactive Controls
Active & passive fire protection
Access & escape route diversity Access to scene
Emergency response by site and external authorities

Summary

- Although flare and relief systems have been the underlying cause for major incidents they have not received the right attention. They are a critical safety feature hence it is vital that they are efficiently managed and maintained.

- Evergreening proposes a process safety management program to ensure effective and sustainable overpressure protection for all plant equipment.

17

Saudi Aramco: Confidential

أرامكو السعودية
saudi aramco

Questions?

18

Saudi Aramco: Confidential

أرامكو السعودية
saudi aramco

PROCESS SAFETY OF KEY INTEREST IN INSURANCE RISK ASSESSMENT SURVEYS

10 Oct 2017

CCPS MIDDLE EAST PROCESS SAFETY CONFERENCE

Vinod Menon, Haifa Al-Marzooq and S.Kugan
Risk and Reliability Management Section
Operational Excellence Department
BAPCO, Kingdom of Bahrain

AGENDA

- Industry Loss Statistics
- Insurance risk assessment survey
- Key items of focus during survey
- Concluding Remarks

Top cause of business interruption losses **across all lines of business.**

$ 2m+
Value of the average large BI property insurance claim

2/3rd is due to Fire/Explosion and Machinery Breakdown

Source: Allianz Global Corporate & Specialty: Global Claims Review 2015

Top cause of business interruption losses **in Energy Sector**

$ 4m+
Value of the average large BI property insurance claim

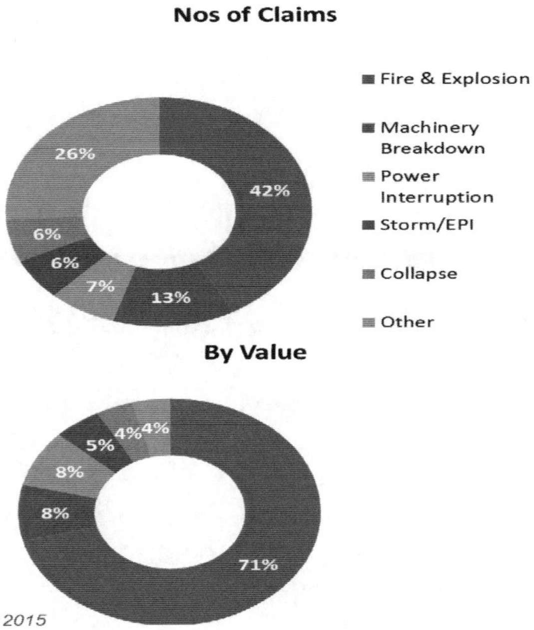

Source: Allianz Global Corporate & Specialty: Global Claims Review 2015

- Energy related Companies (Oil & Gas, Refining, Petrochemical, Power) have complex risk maps and are exposed to a number of Operational risks
- Examples of operational risks relevant to the Energy Industry are
  - Fire / Explosion / Toxic release from a hydrocarbon loss of containment incident
  - Breakdown of a process equipment, e.g. Centrifugal Compressor
- These risks could result in an insurance claim from:
  - Property Damage – Physical Damage and Business Interruption.
  - Third Party Liability claims

- Given the size and complexity of the program, typically, Energy Industry Insurance covers are placed through International Reinsurers
- Also, most insurance placements are done through a major international reinsurance broker
- The Risk Assessment survey is carried out (typically once every two years) by the Company's appointed broker and representatives from the lead reinsurers
- The survey report is very comprehensive and provides details of the risk exposures and loss control features of a particular site, which enables an underwriter to arrive at informed decisions about the 'transfer of risk'

- The survey normally include site visits and meetings with all operational areas
- The surveyors also come up with “recommendations” for risk improvement for the Company to consider and implement. The response to the recommendations are closely monitored by the underwriters
- The team carrying out the survey are mostly from the Engineering community, who typically bring in their knowledge and experience (e.g. benchmark, best practice etc.) through their work on a number of surveys

- Insurance Risk Surveys in Energy Industry lays a lot of emphasis on process safety, which is a significant aspect in most operational incidents
- By engaging with the Insurance community, the Energy Industry can add value to their risk management process, which can improve their risk profile and eventually improve their safety / loss record

Bapco

Insurance Market Order of Priority

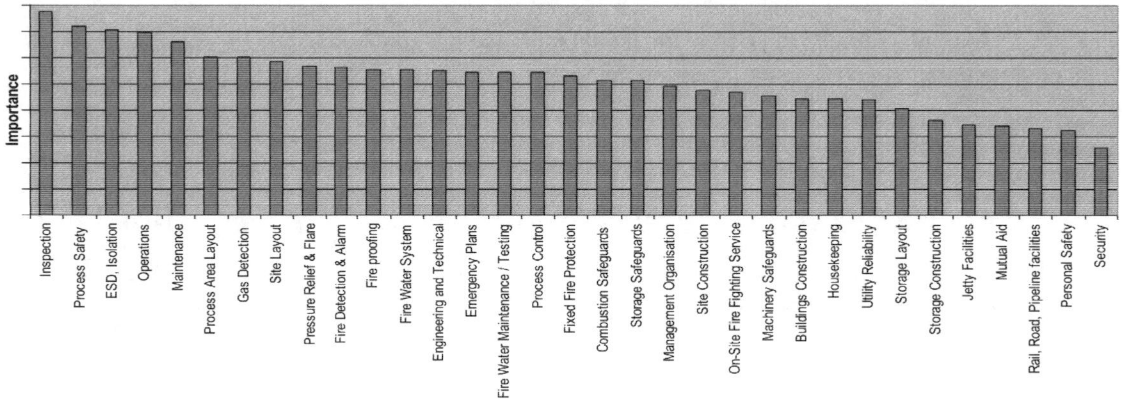

Source: Marsh & McLennan Companies

04 October 2017 Always Improve 9

Bapco

- **From an Operational Risk perspective, following are the areas of focus by the insurance survey community**
  - Mechanical integrity failure due to corrosion
  - Operating practices and procedures
  - Hazard identification studies
  - Permit to Work and Equipment isolation
  - Management of Change
  - Remote Operated Isolation valves
  - Recent Incidents
  - New Projects

Always Improve 10

Bapco

- **Also on the radar are**
  - Safe venting of bonnets of bellows type relief valves
  - Use of long bolt flanges
  - Fire and gas detectors
  - Fire water pumps and annual testing to NFPA requirements
  - P&IDs/ Control of changes
  - Fire proofing of structures

Always Improve 11

Bapco

- **Mechanical integrity failure due to corrosion**
  - Most common cause of losses in the energy sector

54% **Of major losses were due to mechanical integrity failure due to corrosion**

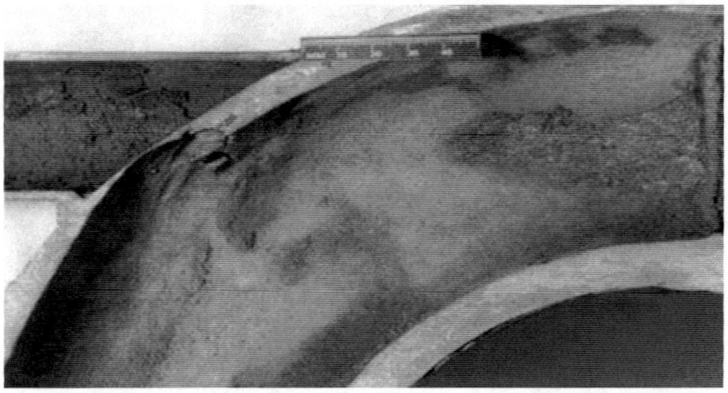

65%	**Internal corrosion of process piping**
13%	**Corrosion of static equipment**
22%	**Incorrect or defective materials**

Source: Key Information Guidelines for Oil, Gas & Petrochemical Risk Engineering Survey Reports Ron Jarvis (Swiss Re London) and Andy Goddard (Talbot Syndicate London)

Always Improve 12

Bapco

- **Mechanical integrity failure due to corrosion**; focus on
  - Inspection program for internal corrosion of piping
  - Positive material identification during construction and maintenance
  - Inspection of locations that are hard to access
  - Inspection deferments
  - External corrosion under insulation
  - Adequacy of resources to meet the inspection program requirements
  - Qualification of Inspectors
  - Organizational hierarchy

Always Improve 13

Bapco

- **Operating practices and procedures;** focus on
  - Management of bypassing and disabling of safety critical instruments (trips and interlocks)
  - Emergency operating procedures (presence, availability, effectiveness, readiness)
  - Staffing levels
  - Shift handover

Always Improve 14

Bapco

- **Operating practices and procedures;** focus on Shift handover

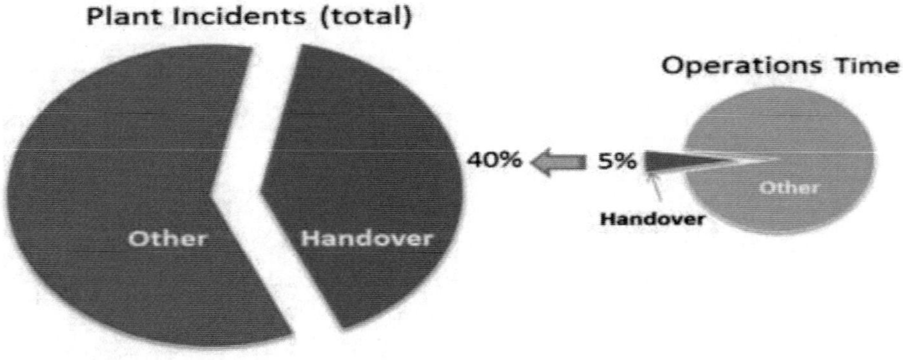

(Source: NPRA 2009 National Safety Conference)

Always Improve 15

Bapco

- **Operating practices and procedures;** focus on Shift handover

- Two way channels for discussion
- One on One interactions; Question: Clarify
- Physical logs -avoiding technical misunderstanding.

(Source: NPRA 2009 National Safety Conference)

Always Improve 16

Bapco

- # Long bolt flanges

  - Also called long bolt valves, sandwich, or wafer valves. Devices such as control valves, butterfly valves and check valves with no integral flanges for bolting to the pipe or vessel flanges, and thus having exposed bolts of >3-inches
  - The long bolts will likely receive direct flame contact (Impingement) if there is a fire in the area
  - Direct flame contact causes the bolts to expand rapidly and lengthen, allowing both gaskets to leak product in <<20minutes. The leaking material then adds fuel to the fire and if under pressure it causes a large spraying fire that results in much more damage

Always Improve 17

Bapco

- **Long bolt flanges;** Action following insurance recommendations

- Identify Long Bolt Flanges
- Assess fire hazard from the contained material and location hazards
- Where possible, replace with normal flange valves
- Reduce risk by wrapping the long bolts with a fire resistant material and/or enclosing the entire assembly with a stainless steel covering

Bapco

Long Bolt Flanges with stainless steel covering

Always Improve 21

Bapco

- **Bellows Type Relief Valve-safe venting of bonnets**
  - Functionality of a bellows type Relief Valve (RV) requires that the RV bonnet operates at atmospheric pressure at all times; hence bonnet is provide with a vent hole
  - A failure of the bellows often results in leakage of fluid from the downstream side of the RV into the bonnet and out of the bonnet vent. Therefore API RP520 requires that the bonnet is vented to a safe location
  - Plastic caps are provided on the bonnet vent hole during transportation
  - Bug screen may be provided for the vent hole

Always Improve 22

Bapco

RV with ½" NPT bonnet vent

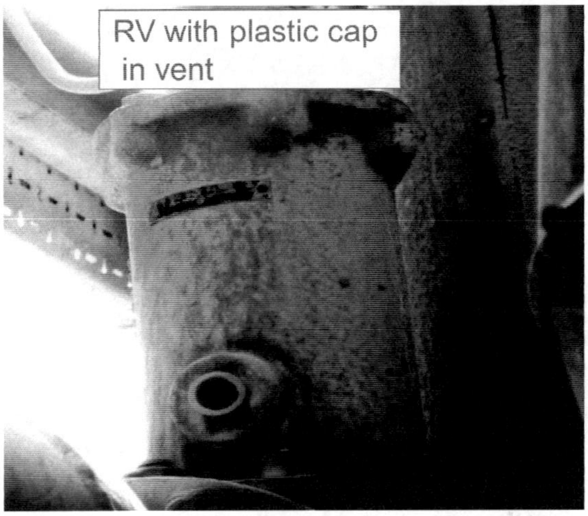
RV with plastic cap in vent

Bapco

RV with vented plug with bug screen

Bug screen is painted over and blocked

Bapco

- **Remote Operated Isolation valves**
  - ROIV or EIBV (Emergency Isolation Block Valve) is dedicated to the purpose of isolating large inventories of flammable or toxic material from sources or equipment whose relative likelihood of significant leakage is high
    - Prevent escalating of a LOC event
  - Insurance Focus
    - Site has appropriate standards and requirement for installing in new units and retrospectively in existing units
    - ROIVs are maintained to the prescribed standards

Always Improve 25

Bapco

- **ROIV/EIBV Bapco experience**
  - Standard specifies requirement basis products characteristics and inventory
    - Light / Heavy components present
    - Operating temperature above/below Flash Point
  - Use of Manual or Remote Operated valves
  - Mandatory for new units
  - Existing units-evaluated during the Hazop process ; investment decisions based on risk analysis/remanant life and feasibility

Always Improve 26

Bapco

- **ROIV/EIBV Bapco experience**
  - Fireproofing/Fire-resistant specifications for Valve, Actuator and cabling
    - Fire resistance time duration of 30 minutes
    - Remote actuation switch at a safe location in the field
  - Issues
    - Actuator damage from process temperature
    - Inadequacy of fireproofing after maintenance

Always Improve 27

Bapco

Concluding Remarks

- The energy industry has become safer over the past several years, as companies have come to realize the high costs associated with accidents.
- However, when accidents occur, insurance claims are generally higher than in the past, as a result of the utilization of more complex technology.
- It is our experience that process safety on site has benefitted from the recommendations that come out of the Insurance Risk surveys. The insurance 'buying' process should consider the services provided by the brokers and underwriters to help in improving the risk profile.

Always Improve 28

Bapco

Concluding Remarks

- Survey findings and the follow up on recommendations have enabled us to have a relook and close gaps in our systems and also is a reassurance, that we are treading the right path on our PSM journey.
- We believe the insurance survey is a beneficial process - sites benefit from improved process safety which can contribute to improved risk profile and lower premium spend.
- The Energy Industry should positively engage with their insurance partners in order to achieve the benefits out of the survey.

Bapco

THANK YOU

HAZARD RECOGNITION PROCESS ENHANCEMENT

(HAZARDS IDENTIFICATION IMPROVEMENT)

Eng. Ibrahim Abdullah Al-Jamaan

Process Safety Leader

CHEMISTRY THAT MATTERS™

CONTENTS :-

- Hazard Vs. Risk Definition
- Hazard Identification Methodologies
- Risk Management starts with Hazard Identification
- Hazard Recognition 2015 Vs. 2017
- Success Factors
- Success Secret

HAZARD IS ....

- A source of potential harm, or a situation with a potential for causing harm, in terms for human injury, damage to health, property, environment, and other things of value or some combination of these.

No. 2

RISK

The effect of uncertainty on objectives.

NOTE 1: An effect is a deviation from the expected — positive and/or negative.
NOTE 2: Objectives can have different aspects (such as financial, health and safety, and environmental goals) and can apply at different levels (such as strategic, organization-wide, project, product and process).
NOTE 3: Risk is often characterized by reference to potential events and consequences, or a combination of these.
NOTE 4: Risk is often expressed in terms of a combination of the consequences of an event (including changes in circumstances) and the associated likelihood of occurrence.
NOTE 5: Uncertainty is the state, even partial, of deficiency of information related to, understanding or knowledge of an event, its consequence, or likelihood.

No. 3

HAZARD IDENTIFICATION

The process of recognizing that a hazard exists and defining its characteristics

No. 4

METHODS & SOURCES

No. 5

SABIC PRACTICE : HAZARD IDENTIFICATION ELEMENTS

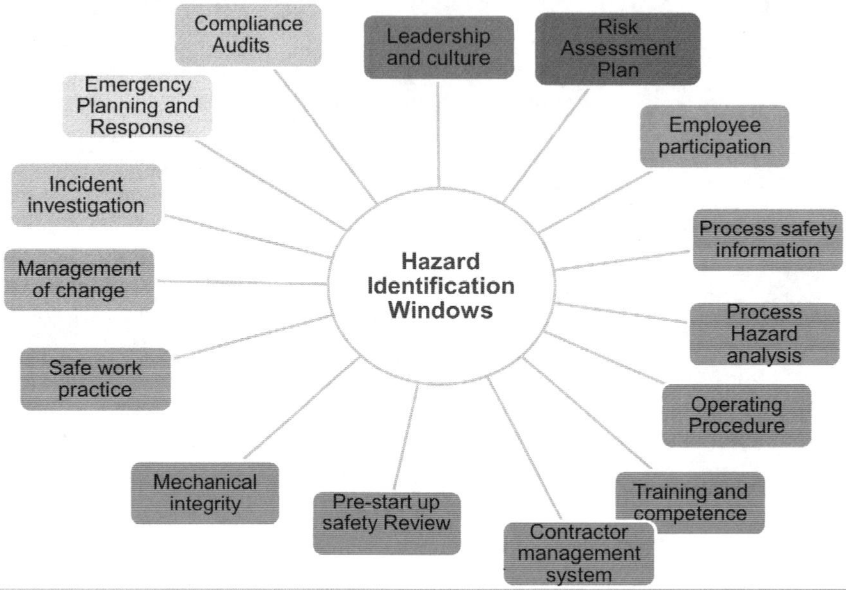

No. 6

HAZARD ANALYSIS:-

- The process of recognizing hazards that may arise from a system or its environment, documenting their unwanted consequences and analyzing their potential causes.

No. 7

RISK ANALYSIS

The systematic use of information to identify hazards and estimate the chance for, and severity of, injury or loss to individuals or populations, property, the environment, or other things of value.

No. 8

RISK ASSESSMENT

Overall process of identification of hazards, evaluation of risks and determination of appropriate measures to control risks.

No. 9

HAZARD CLASS

Nature of the physical, health or environmental hazard

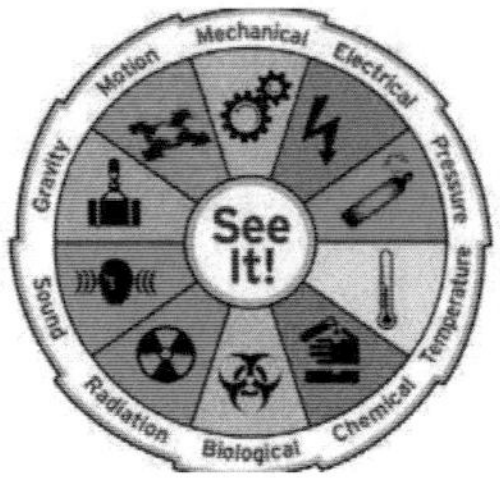

No. 10

HAZARD CLASSIFICATION

- The precise identification of the hazard(s) of product (substance or mixture) or raw material by assigning minimum hazard symbols, hazard statements and precautionary statements based on defined national and international legislation.

No. 11

HAZARD RECOGNITION PROCEDURES

- The process of identifying, evaluating and monitoring hazards in the work environment .

HAZARDS TYPES :-

A common way to classify hazards is by category:
biological - bacteria, viruses, insects, plants, birds, animals, and humans, etc.,
chemical - depends on the physical, chemical and toxic properties of the chemical,
ergonomic - repetitive movements, improper set up of workstation, etc.,
physical - radiation, magnetic fields, pressure extremes (high pressure or vacuum), noise, etc.,
psychosocial - stress, violence, etc.,
safety - slipping/tripping hazards, inappropriate machine guarding, equipment malfunctions or breakdowns.

EXAMPLES:-

Workplace hazards can come from a wide range of sources. General examples include any substance, material, process, practice, etc. that has the ability to cause harm or adverse health effect to a person or property. See Table 1.
Workplace hazards also include practices or conditions that release uncontrolled energy like:
•an object that could fall from a height (potential or gravitational energy),
•a run-away chemical reaction (chemical energy),
•the release of compressed gas or steam (pressure; high temperature),
•entanglement of hair or clothing in rotating equipment (kinetic energy), or
•contact with electrodes of a battery or capacitor (electrical energy).

Table 1
Examples of Hazards and Their Effects

Workplace Hazard	Example of Hazard	Example of Harm Caused
Thing	Knife	Cut
Substance	Benzene	Leukemia
Material	Mycobacterium tuberculosis	Tuberculosis
Source of Energy	Electricity	Shock, electrocution
Condition	Wet floor	Slips, falls
Process	Welding	Metal fume fever

HAZARD IDENTIFICATION STRATEGY

No. 16

HAZARD IDENTIFICATION OBJECTIVE

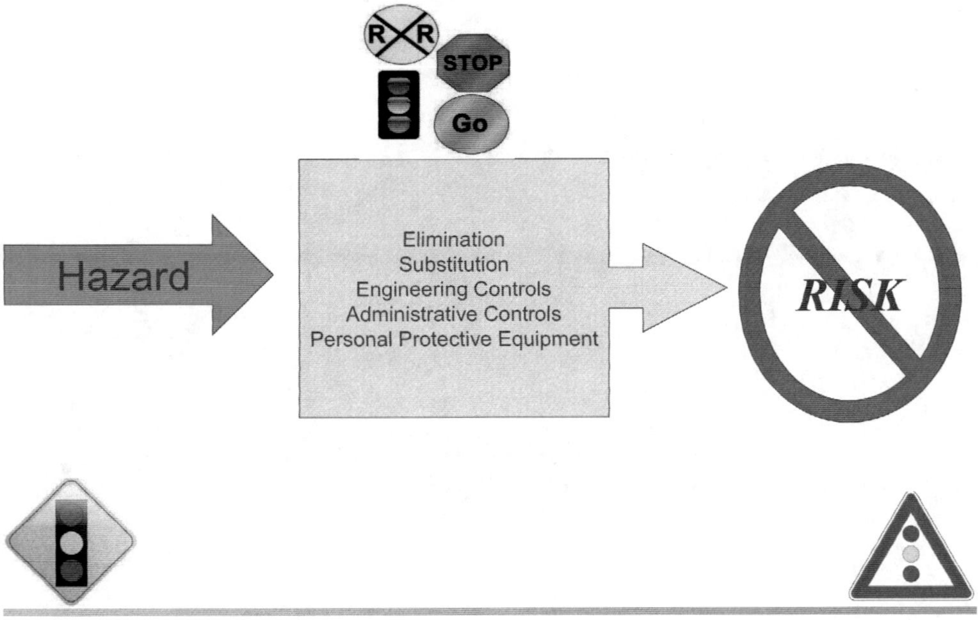

No. 17

HAZARD IDENTIFICATION OBJECTIVE

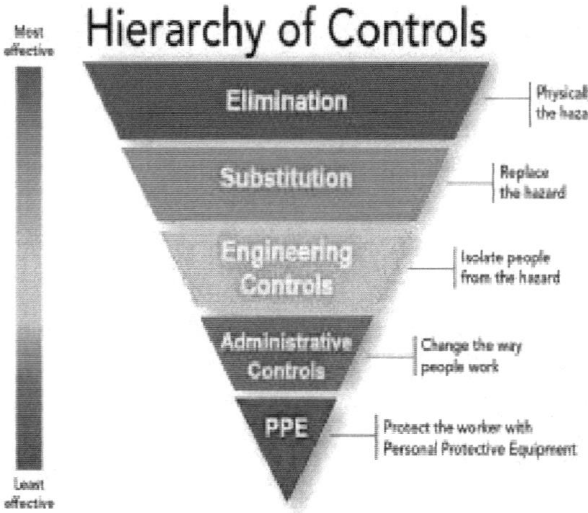

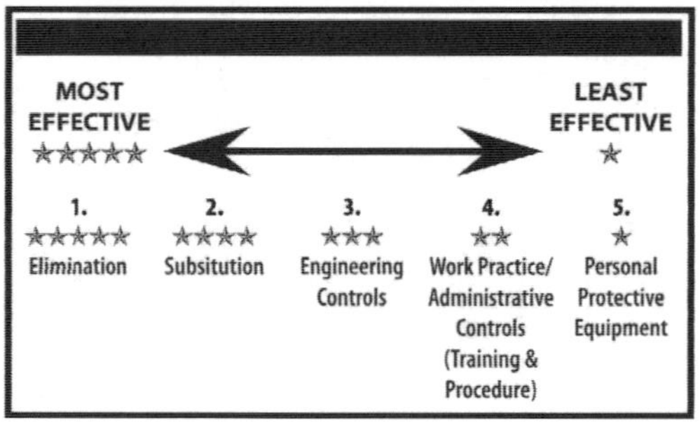

No. 18

RISK ASSESSMENT FLOWCHART

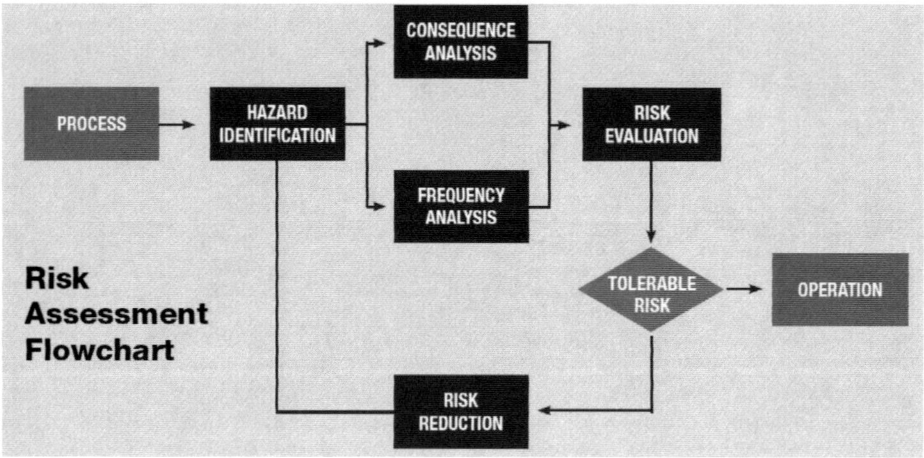

No. 19

PROCESS SAFETY IS OUR CORE VALUE

Process Safety is the Business Chain that contains :-

- Hazard Identification
- Risk Assessment
- Risk Management & Control
- Stewardship

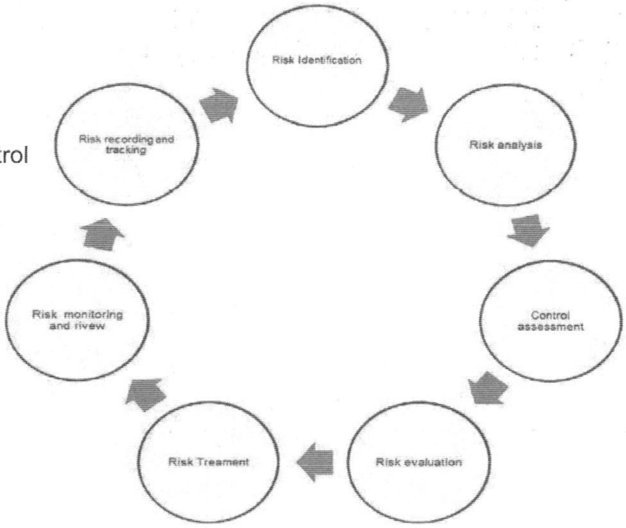

No. 20

ONLY AS GOOD AS THE WEAKEST IN YOUR CHAIN

No. 21

PROCESS SAFETY PYRAMID

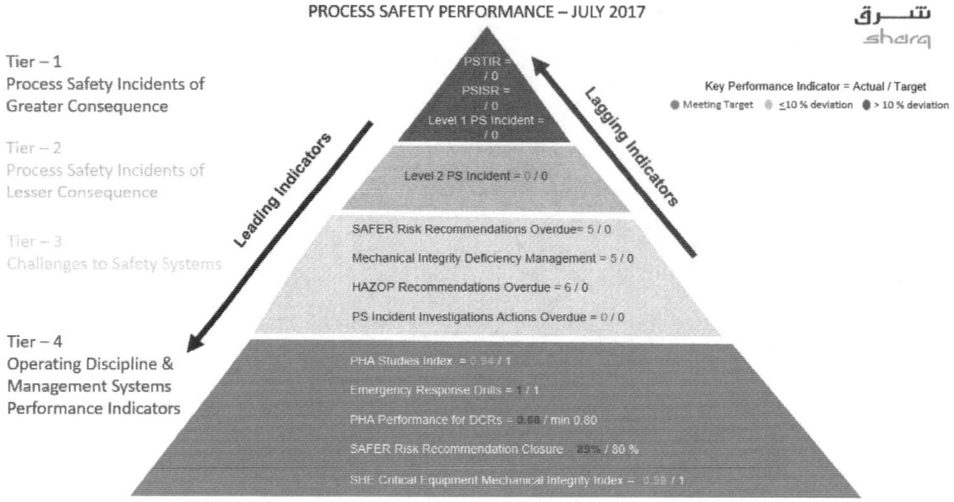

No. 22

SCORE CARDS

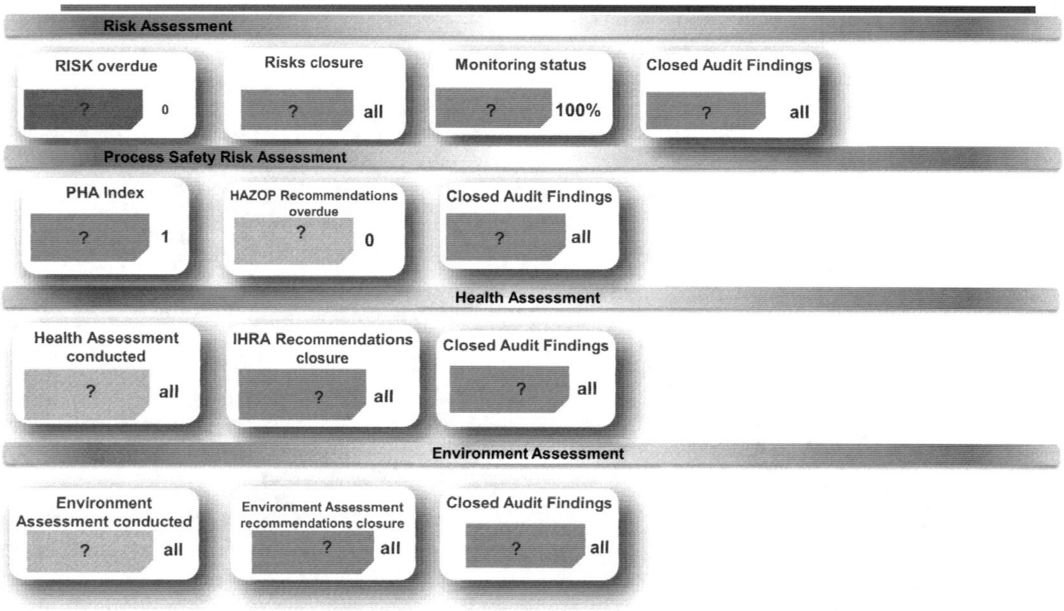

No. 23

sharq

RISKS ANALYSIS (example)

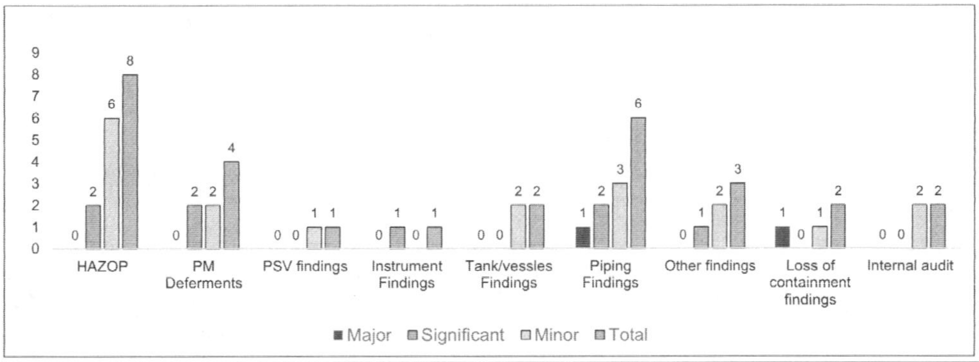

Risk Type	Major	Significant	Minor	Total
HAZOP	0	2	6	8
PM Deferments	0	2	2	4
PSV findings	0	0	1	1
Instrument Findings	0	2	0	2
Tank/ vessels Findings	0	0	2	2
Piping Findings	1	2	3	6
OTHER findings	0	1	2	3
Loss of containment findings	1	0	1	2
Internal Audi findings	0	0	2	2

sabic

Risk Wise Identification Summary

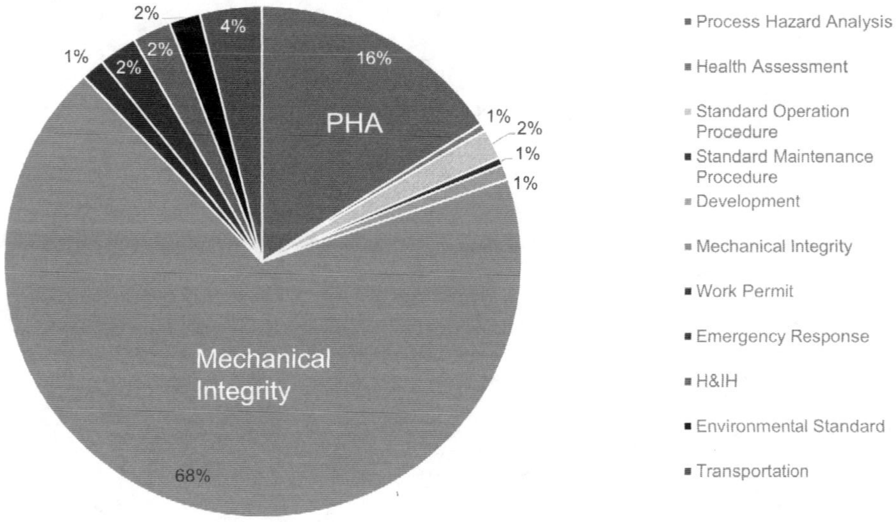

BRADLEY CURVE

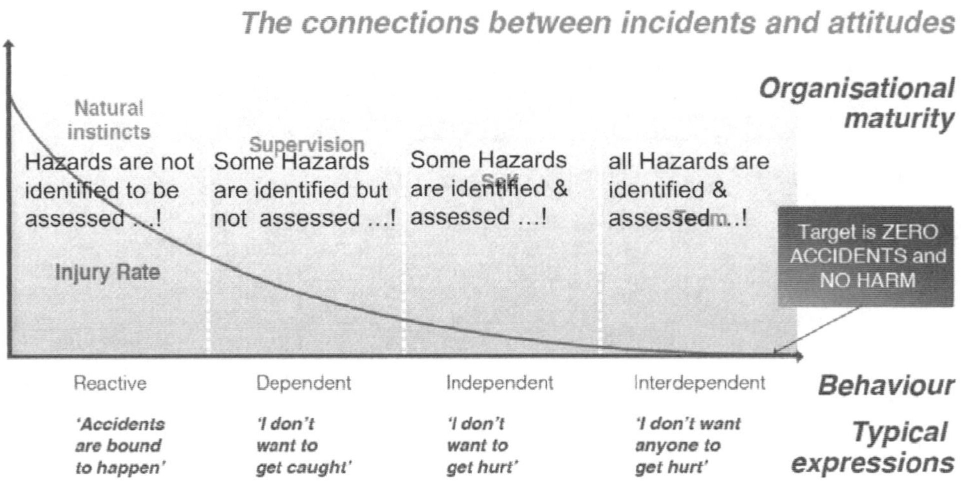

No. 26

BRADLEY CURVE

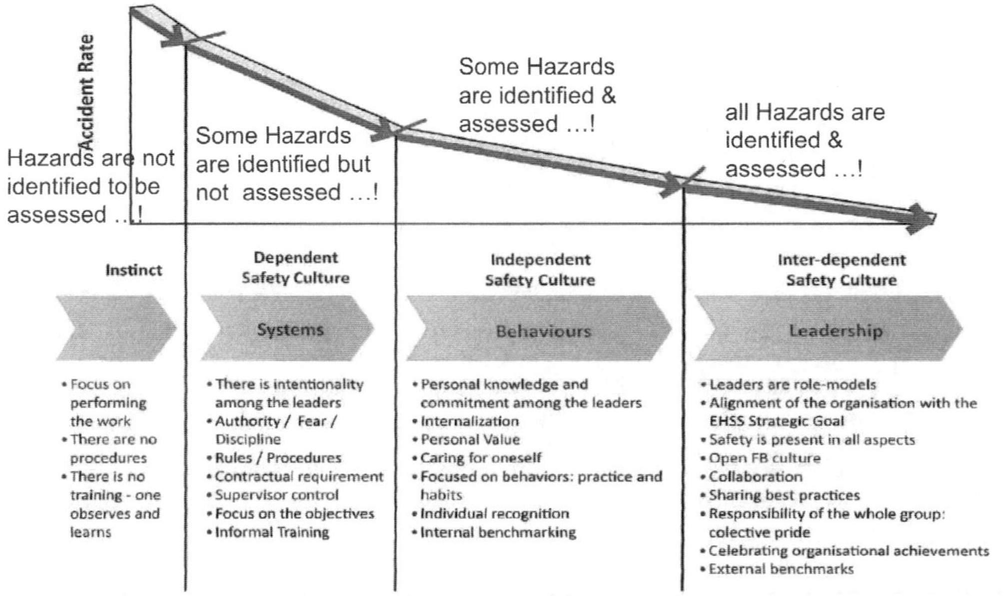

No. 27

سابك
sabic

MAJOR ACOMPLISHMENTS

1 Brand Expression & Guidelines
- Establish Process safety section and Develop PHA leader
- Establish Process safety function

2 Risk Assessment Program
- Improved Risk identification and quality
- Risk Identification and Assessment

3 Training and Awareness
- Awareness provided to Senior management and employees
- Provide training awareness to the employee and leaders

4. Risk Management Implementation
- Developed Q pulse Risk Form, Performance monitoring through PS KPIS
- Implement and monitor Sisk Management program

5. Process Safety Culture
- Arranged Process safety campaign to start cultural change
- Organization wide awareness to build process safety culture

No. 28

سابك
sabic

MAJOR ACOMPLISHMENTS (APPROXIMATED NUMBERS)

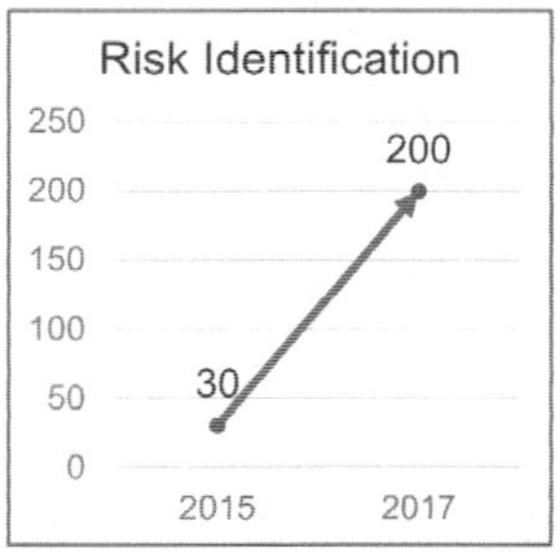

No. 29

PROCESS SAFETY ACCOMPLISHMENT

Engagement with Sabic Process Safety Committee

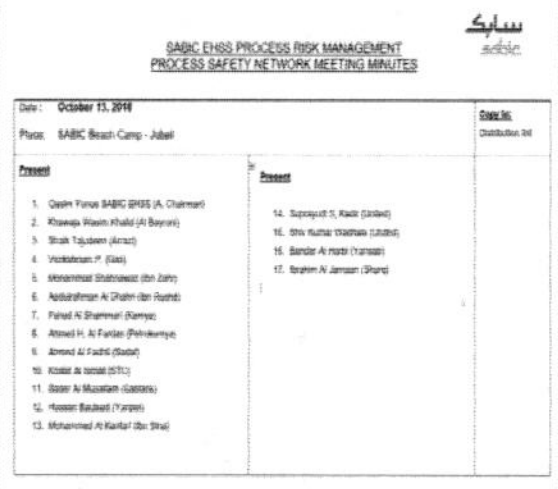
SABIC EHSS PROCESS RISK MANAGEMENT
PROCESS SAFETY NETWORK MEETING MINUTES

Process Safety & Risk Management Visit

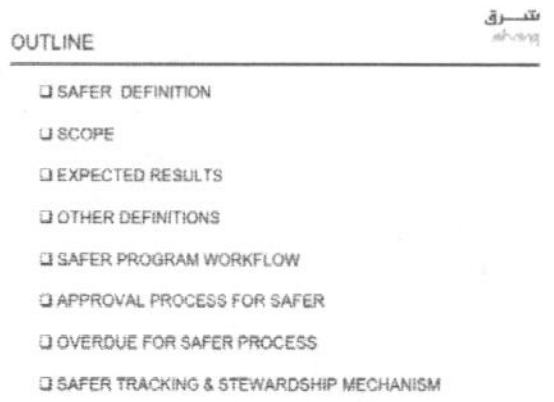
OUTLINE

- SAFER DEFINITION
- SCOPE
- EXPECTED RESULTS
- OTHER DEFINITIONS
- SAFER PROGRAM WORKFLOW
- APPROVAL PROCESS FOR SAFER
- OVERDUE FOR SAFER PROCESS
- SAFER TRACKING & STEWARDSHIP MECHANISM

Develop and Implement PS KPI's

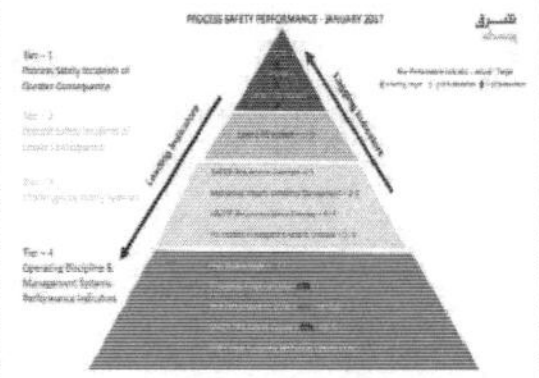

Launching Process Safety walkthrough

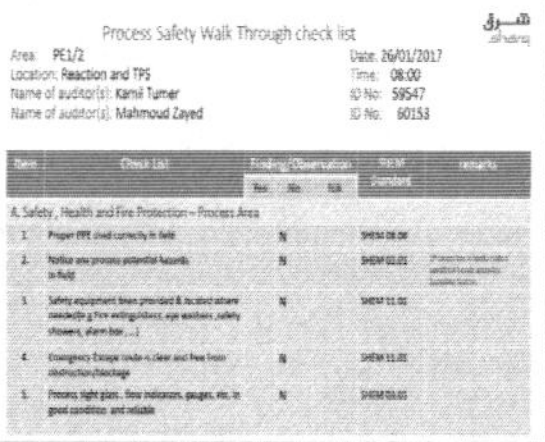
Process Safety Walk Through check list

Risk Management Intensive Training

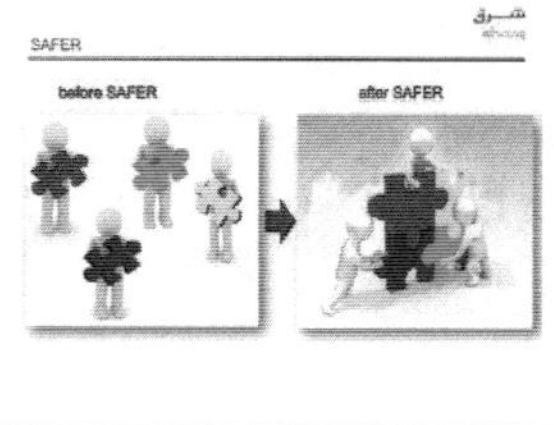

Participation in 3rd CCPS Global Summit on Process Safety

PROCESS SAFETY ACCOMPLISHMENT

Certified PHA Leaders
2014:3
2015:+8
2017: +11

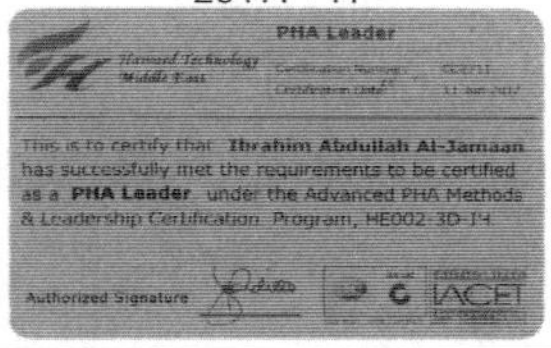

EHS Critical list verified and identified during Re-do-Hazop

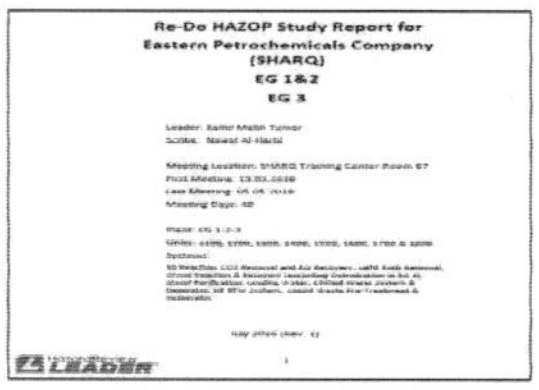
Re-Do HAZOP Study Report for Eastern Petrochemicals Company (SHARQ)
EG 1&2
EG 3

Process safety Portal

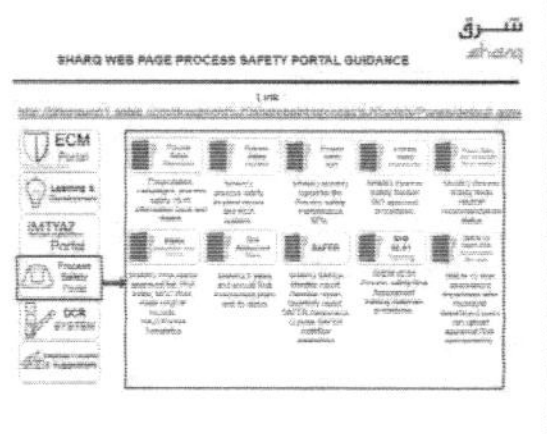

2017 PATH FORWARD

Process Safety Campaign

Process Safety Intensive Training

- Gain an understanding of the difference between occupational and process safety.
- Come to the realisation that all process safety incidents are preventable.
- Become familiar with the tools that enable process safety.
- Become aware of the required behaviour to deploy process safety in the organisation.
- Become familiar with the tools that will drive the required behaviour

Risk Assessment Quality Enhancement

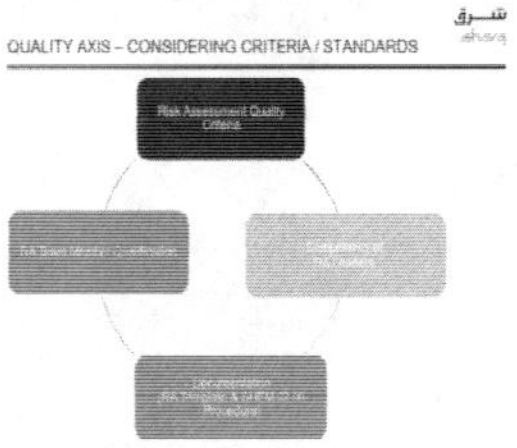

Process Safety Engagement for front liner

S.N.	Volunteers	Department	(Part Time)
1	Ghadban-Al, Abdullah Khalid	CHMD	Vendor Co-Leader
2	Beshi-Al, Abdulaziz Turki	EGMD	Vendor Co-Leader
3	Rabeah-Al , Qassim	CHMD	13th of February
4	Mutairi-Al , Nawaf N	CHMD	13th of February
5	Harith Naif	CMD	14th of February
6	Tammar-Al , Mohammed A	EG 1&2	13th of February
7	Ali Marshood Alshammari	EG 1&2	13th of February
8	Mohammed Fahad Alqahtani	EG 1&2	14th of February
9	Khalid Yahya Alabdali	EG 1&2	14th of February
10	Fahad Jamal Aldossary	EG 1&2	15th of February
11	Abdullah Salem Al-jaloud	EG 1&2	15th of February
12	Zahrani-Al, Nawaf Ahmed	EGMD	14th of February
13	Harbi-Al, Saud Ajeeb	EGMD	14th of February
14	Otaibi-Al, Mutlaq H	EGMD	15th of February
15	Mowaid-Al, Ali Talal	O&UMD	15th of February
16	Ibrahim Al-Hussain	PE 3&4	13th of February
17	Abdullah-Al , Salah	PMD	16th of February
18	Zahrani-Al , Rashid A	SD	15th of February

Re-Do-Hazop for Sharq Plants as per Sharq Risk Assessment Plan

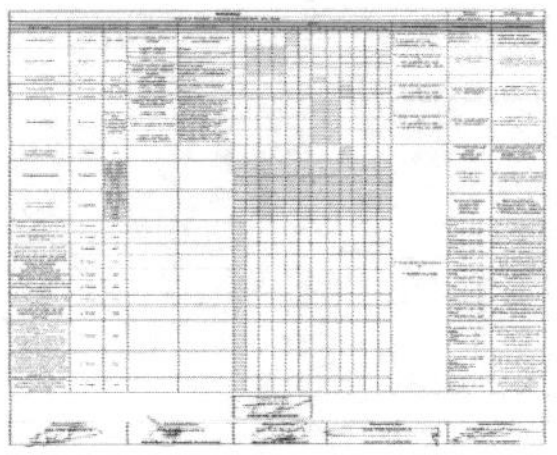

RISK Assessment training

2017 PATH FORWARD

Develop Shift hand over and logbook processes – (Good Practice)

2017

Critical operation limits SOE & COP

2017

2017 Process Safety survey

Process Safety Awareness for front liner

- ✓ Review PSL 1
- ✓ Overview on process safety performance indicators
- ✓ LOPC & Threshold quantity
- ✓ Walkthrough finding discussion
- ✓ Case study & Lesson learned

PHA Leaders Workshop

Process Safety Development Plan

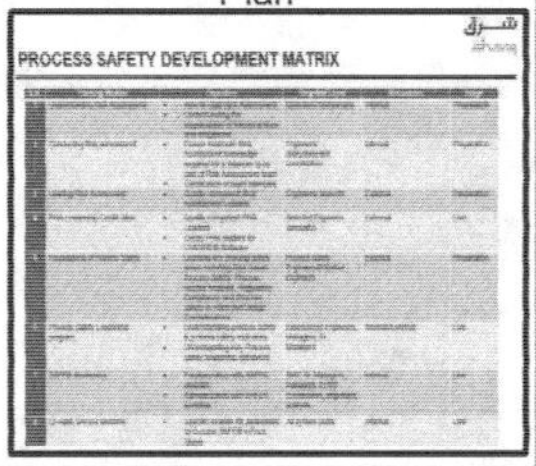

JOURNEY CONTINUES

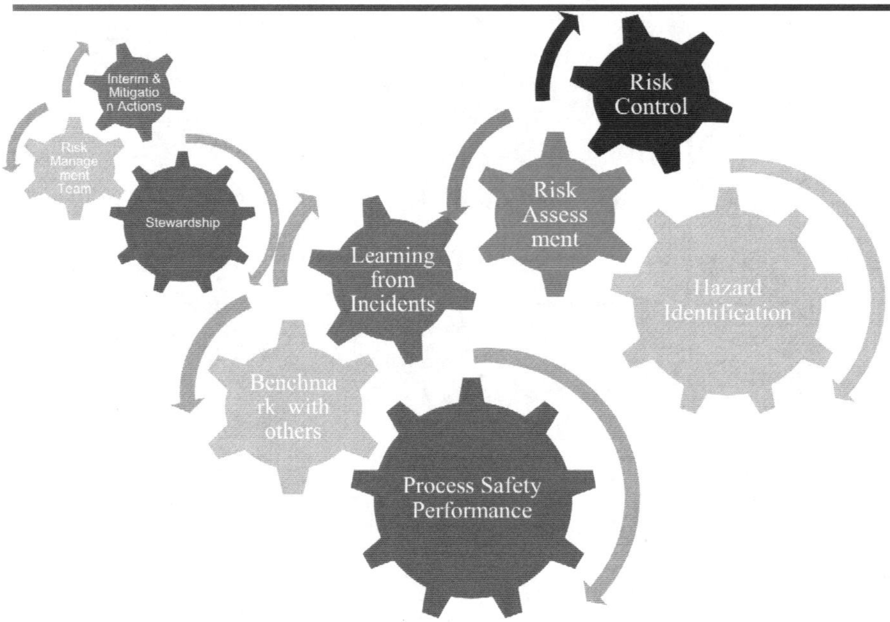

No. 34

CONCLUSION

Hazard Identification is the limiting reagent for the Risk Management …

- if it is enhanced, the other Gears will be enhanced …

- if it is weak; the complete chain will break …

- if it fails, the whole system will fail …

Secret:-

"Engagement" is the secret word to achieve the Process Safety Excellence that drives all individuals to think & consider any identified Hazard in the whole organization.

No. 35

DISCLAIMER

INFORMATION AND RECOMMENDATIONS CONTAINED IN THIS DOCUMENT ARE GIVEN IN GOOD FAITH. HOWEVER, SABIC AND ITS AFFILIATES MAKE NO EXPRESS OR IMPLIED REPRESENTATION, WARRANTY OR GUARANTEE (i) THAT ANY RESULTS DESCRIBED IN THIS DOCUMENT WILL BE OBTAINED UNDER END-USE CONDITIONS. SABIC AND ITS AFFILIATES SHALL NOT BE RESPONSIBLE FOR ANY LOSS RESULTING FROM ANY USE OF RECOMMENDATIONS OR INFORMATION DESCRIBED IN THIS DOCUMENT. Each user is responsible for making its own determination as to the suitability of any information or recommendations herein for the user's particular use through appropriate end-use and other testing and analysis. Nothing in this document or any oral statement shall be deemed to alter or waive any provision of SABIC's or its affiliates' Standard Conditions of Sale or this Disclaimer, unless it is specifically agreed to in a writing signed by SABIC or its affiliates. Statements by SABIC or its affiliates concerning a possible use of any material, product, service, process, method, or design do not, are not intended to, and should not be construed to grant any license under any patent or other intellectual property right of SABIC or its affiliates or as a recommendation for the use of any material, product, service, process, method, or design in a manner that infringes any patent or other intellectual property right.

SABIC and brands marked with ™ are trademarks of SABIC or its subsidiaries or affiliates, unless otherwise noted.

Any brands, products or services of other companies referenced in this document are the trademarks, service marks and/or trade names of their respective holders.

No. 36

Thank you

No. 37

Arthur D Little

eni versalis

A Value-Based Approach for the Evaluation of Investment Options in Oil & Gas Plants

Stefano Milanese

Bahrain International Exhibition & Convention Centre

Manama, Bahrain

Tuesday, 10th October 2017

Introduction – Our Partnership with Eni versalis

eni versalis

Arthur D. Little and versalis have established a collaboration aimed at developing a structured multi-year Process Safety Management program

Eni Versalis

- Versalis is a leading multinational Chemical and Petrochemical Company part of the eni Group
- Versalis owns Production Plants all over Europe
- 5.7 million tons of petrochemical products and polymers in 2016

Arthur D. Little

- As the world's first consultancy, Arthur D. Little (ADL) has been at the forefront of innovation for more than 130 years
- Arthur D. Little stands out for its Side-by-side approach with Clients at all levels
- We leverages on a solid experience with companies and governments in complex business and social contexts

Process Safety Management Program

Process Safety Management is a strategic process related to the business performance and in particular to asset integrity, production continuity, safety of personnel, contractors and population, environmental protection and reputation of the Company

Arthur D Little 2

In the recent past years, the Oil & Gas context faced several changes due to limited resources, assets ageing and stringent standards…

Oil & Gas Issues

Oil prices reduction: after the 60% drop in the second semester of 2014, the Brent price has not gone above US$60 since

Assets ageing: many assets are operated beyond their originally designed lifespan

Stricter Industry Standards and **Regulations**, required by Companies' stakeholders

Challenges

- Oil & Gas Companies face now a choice regarding old assets and compliance with Industry Standards and Regulations:
  - complete decommissioning / shut down
  - large investments into **plant upgrade** and/or **revamping**

… making it essential to adopt structured approaches to support Management decision-making process in **selecting and prioritizing investment options,** so as to create value in the future

A Safety Integrity Level (SIL) initiative was executed as part of the overall PSM program in order to ensure effective management of process risks

SIL Analysis

- Safety Integrity Level analysis assesses risks due to the failure of **Safety Instrumented Functions (SIF)**
- Each SIF is assigned a **Safety Integrity Level (SIL),** which represents an average Probability of Failure on Demand

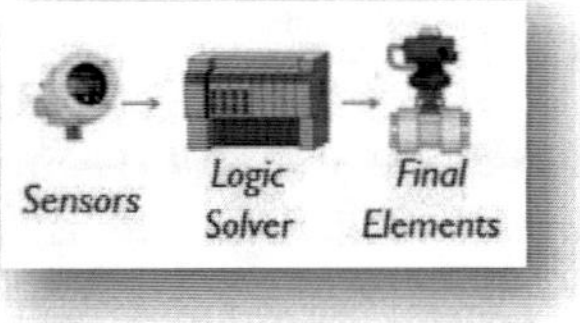

1. **SIL Allocation** identified the reliability requirements for each SIF
2. **SIL Verification** determined the current reliability provided by the installed components
3. **SIL Optimisation** was applied to SIFs presenting a reliability gap (i.e. SIL allocated greater than SIL verified), determining the improvement actions needed to close each gap

The SIL initiative involved 14 Versalis' Petrochemical Plants: 9 in Italy, 2 in UK e 1 respectively in France, Germany and Hungary

Overall Results of SIL Analysis

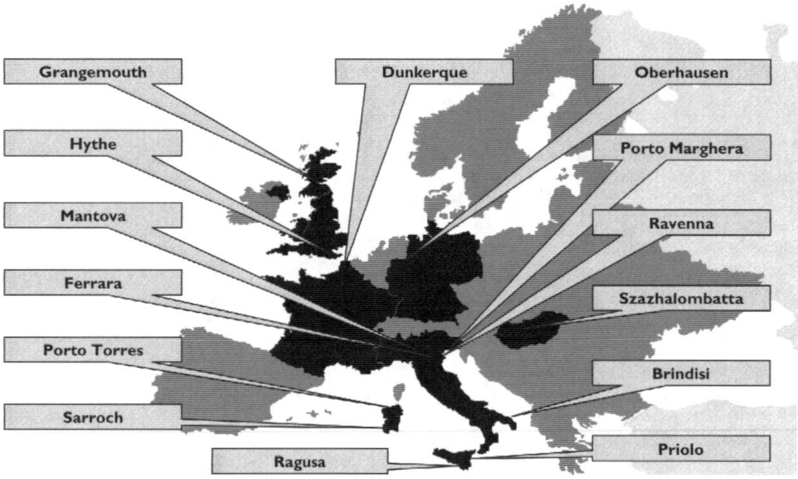

1. **SIL Allocation**
   - 5500 SIFs analysed
2. **SIL Verification**
   - 3000 SIFs analysed
   - 1700 SIFs in Gap
3. **SIL Optimisation**
   - 1500 SIFs analysed
   - 1000 SIFs in Gap
4. **Cost Benefit Analysis**
   - Selection of the most feasible improvement actions to close the reliability Gaps

Arthur D Little 5

Improvement actions needed to close the reliability gaps concern both technological and organizational measures

Technological and hardware changes

- Integration or replacement of installed components with **more reliable devices**
- Modification of existing **protection layers** or implementation of additional ones
- Installation of **new alarms** granting enough time for the operator to perform remedial actions

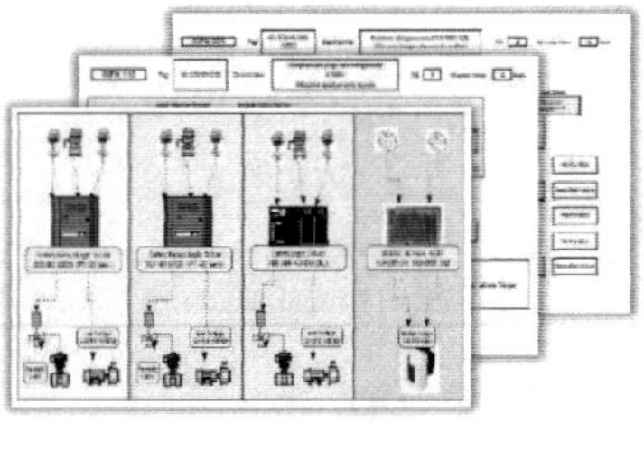

Organizational / procedural measures

- Increase of **maintenance frequency** of installed components
- Revision or introduction of **operating procedures**, **emergency plan** (internal or external)

Arthur D Little 6

For each improvement action, required investments involve both the implementation and the maintenance of adopted measures

$$\textbf{Annual cost of the improvement action} = \frac{\text{CAPEX}}{\text{Depreciation period}} + \text{OPEX}\ (€/\mathit{year})$$

- Following the identification of the improvement actions for each SIF, costs have been evaluated considering:
  - The cost for the **action implementation**, CAPEX (CAPital EXpenditure) (e.g. new safety equipment, new certified instrumentation)
  - The cost related to **maintenance** of the adopted measures, OPEX (OPerating EXpenditure); (e.g. periodic tests / safety studies, equipment maintenance)
  - **Synergies** arising from the implementation of more than one specific improvement action (e.g. the cost of a new Logic Solver which can be distributed on every SIF which will benefit from it)

Note: the depreciation period for the CAPEX is spread over10 years in equal instalments

The application of improvement actions generates benefits in terms of prevention of impacts affecting production, safety and environment

Benefits considered…

… *for Production* [a)]

… *for people Safety*

… *for Environment and Reputation*

Steps for Benefits evaluation

01 **Current risk (R_C)** calculation (before the investment) considering the SIL Verified

$$R_C = \frac{\mathrm{W \cdot M}}{10^{[\mathrm{SIL_V} + \sum_i IPL_i]}} \cdot \varphi_{\ \mathrm{EF}}\ (€/\mathrm{year})$$

02 **Mitigated Risk (R_M)** Calculation (after the investment) evaluating the implemented actions

$$\mathrm{R}_M = \frac{\mathrm{W \cdot M}}{10^{[\mathrm{SIL_V} + \sum_i IPL_i]}} \cdot \varphi_{\ \mathrm{EF}} \cdot \alpha\ (€/\mathrm{year})$$

03 Evaluation of **Benefits,** stemming from the risk reduction

$$\text{Benefits} = \mathrm{R}_C - R_M = \frac{\mathrm{W \cdot M}}{10^{\sum_i IPL_i}} \cdot \varphi_{\ \mathrm{EF}} \cdot [1 - \alpha]$$

Note: a) Business interruption and Asset damage

b) **W**: Process upset frequency; **M**: magnitude of the accidental scenario; **SILv**: SIL value verified; **IPL**: independent protection layer; φ_{EF} = risk reduction coefficient given by the enabling factors of the scenario; **α** = risk reduction coefficient given by the improvement action

Cost Benefit Analysis – A value-based approach – Indirect benefits

eni versalis

In addition to the previously benefits, SIL analysis generates significant indirect advantages, not included in the Cost Benefit Analysis

Indirect Benefits

- Mapping of critical SIFs and optimisation of resource allocation
- Optimisation of predictive maintenance planning and replacement of obsolete components
- Certification of the correct application of international standards (IEC 61511 and IEC 61508) through accredited tools
- Encouragement of economies of scale for supplies and maintenance plans
- SIL analysis as proof of solid risk management to external verification bodies (e.g. fire departments, insurance and authorities)
- Improvement of Company's reputation
- Implementation of a Digital System including SIL results, so as to enable a dynamic data analysis, both for Management and operating activities

Arthur D Little 9

Cost Benefit Analysis – A value-based approach – Cost Benefit Ratio (1/3)

eni versalis

To help select and prioritize investments, two quantitative parameters have been determined during the Cost Benefit Analysis

Cost Benefit Ratio (CBR)

$$CBR = \frac{Costs}{Benefits}$$

- It represents a dimensionless indication of the **investment convenience**
- Calculated for **every SIF, Plant** and **Site**
- The lower the CBR, the more justifiable the investments

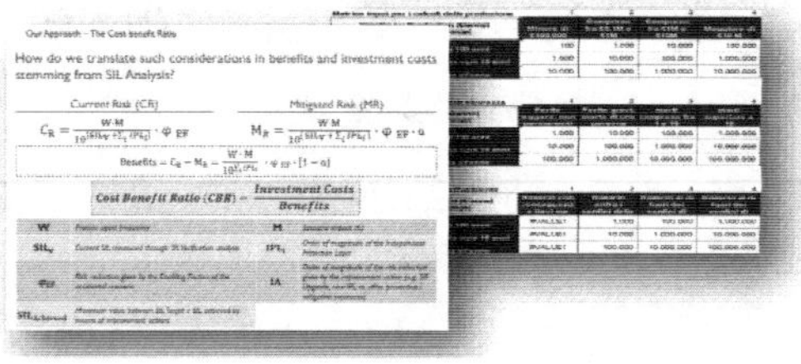

Weighted Cost Benefit Ratio (WCBR)

$$WCBR_{Plant} = CBR_{Plant} * \frac{Plant\ investment}{Site\ Investment}$$

- **It weighs the Plant CBR** considering the fraction of the Site investment required
- Calculated at **plant-wide level**
- Low WCBR indicates that the Plant requires limited investments compared to the overall Site investments

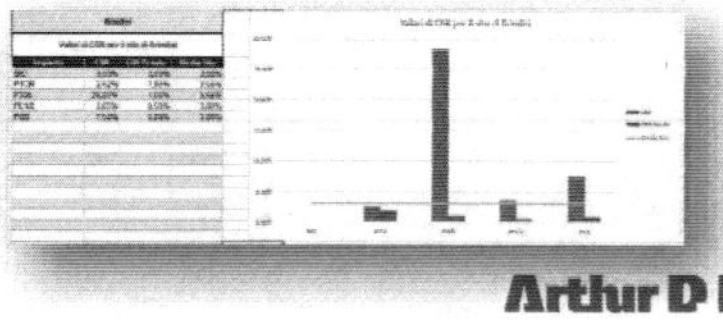

Arthur D Little 10

Cost Benefit Analysis – A value-based approach – Cost Benefit Ratio (3/3)

eni versalis

As shown in the example, taken from Brindisi Site, the Cost Benefits Analysis helps Management compare investments and plan a resource allocation

Illustrative – Brindisi Site

No. SIFs requiring investments	**39**	**2**	**35**	**7**
Plant investment (k€)	**112.7**	**6**	**21.8**	**18.7**
Plant investment (% compared to Site investments)	**71%**	**4%**	**14%**	**12%**
Plant Benefits (k€)	**4,299**	**21.2**	**595.5**	**247.5**

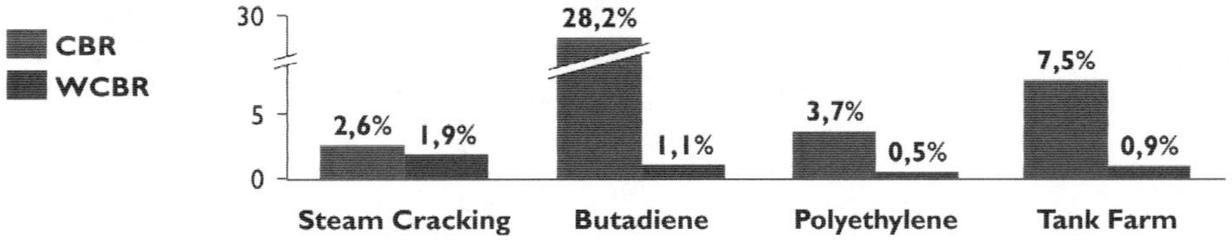

- In the given example, the calculated **WCBR**s are lower than 2%, showing that investments are balanced in all of the four Plants in the Site
- Based on Plant **CBR**s, a reasonable resource allocation would be achieved starting to invest in Steam Cracking and Polyethylene Plants, followed by Storage facility and Butadiene Plant

Arthur D Little 11

Cost Benefit Analysis – A value-based approach – Overall results

eni versalis

Once Resource Allocation Plans are provided by each Site, the Client Headquarter will consequently develop a structured Investment program

Illustrative

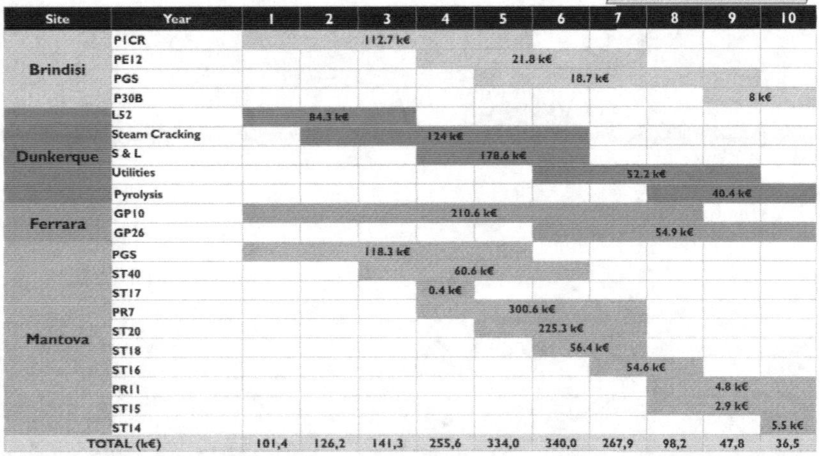

... ensuring an **effective resource deployment** in its Sites and Plants

Arthur D Little 12

Conclusions – The Cost / Benefit Analysis

eni versalis

Arthur D. Little's framework for the design of a Value-Based Approach for the Evaluation of Investment Options in Oil & Gas Plants

Conclusions

- The Cost / Benefit analysis supporting large investments stemming from Process Safety Studies involves several benefits, in particular:

Help with the selection of investments and their prioritization

Define objective criteria to compare costs and benefits at Plant, Site and Company wide level

Facilitate the development of a structured investment Plan over time

Assess the current implementation of specific Process Safety elements inside Sites and Plants

Arthur D Little 13

Arthur D Little

Arthur D. Little has been at the forefront of innovation since 1886. We are an acknowledged thought leader in linking strategy, innovation and transformation in technology-intensive and converging industries.
We navigate our clients through changing business ecosystems to uncover new growth opportunities. We enable our clients to build innovation capabilities and transform their organizations.

Our consultants have strong practical industry experience combined with excellent knowledge of key trends and dynamics. Arthur D. Little is present in the most important business centers around the world. We are proud to serve most of the Fortune 1000 companies, in addition to other leading firms and public sector organizations.

For further information please visit www.adlittle.com.

Contact

Stefano Milanese
Partner – Sustainability & Risk

Arthur D. Little S.p.A.
Corso Monforte 54
20122 Milan
T: +39 02 67 376 228
M: +39 348 733 2226
E: milanese.stefano@adlittle.com

سابك
sabic

RISK REGISTER IMPLEMENTATION AND MANAGEMENT – PRACTICAL CHALLENGES – A CASE STUDY

Mahesh Murthy – Principal Engineer, Process Safety

CHEMISTRY THAT MATTERS™

سابك
sabic

WHAT AM I TALKING TODAY........

- Risk management – general challenges faced by corporations
- Concept of risk register and implementation challenges
- Implementation case study
- Lessons learned

No. 1

RISK MANAGEMENT CHALLENGES

- Poor visibility of various initiatives
- Various interpretation of risk – vocabulary varies
- Anomalies in risk acceptance criteria –Joint Venture entities and Geographies
- Personnel competency
- Normalization of deviation – too close to identify risks !

No. 2

RISK – ONE NAME MANY APPLICATIONS ....

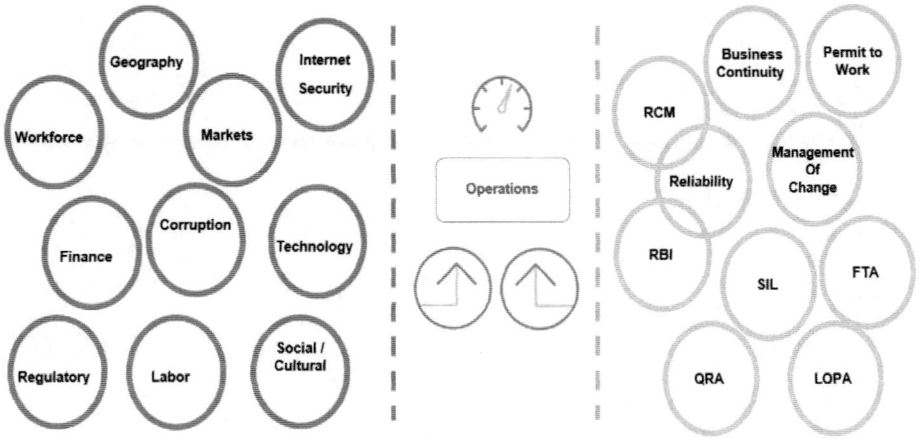

No. 3

RISK MANAGEMENT FRAMEWORK

ISO 31000:2009 – Risk management- Principles and guidelines

ISO/IEC 31010:2009 – Risk Management – Risk Assessment Techniques

ISO Guide 73:2009 – Risk Management- vocabulary

No. 4

RISK REGISTER – WHY DO WE NEED IT ?

- Dashboard visibility
- Tolerable risk acceptance
- Project prioritization and capital allocation
- Upkeeping of adequate controls – Business Assurance
- Plant uptime and business continuity

No. 5

IMPLEMENTATION CHALLENGES

- Leadership awareness
- Process Safety Risk not fully embedded in Enterprise Risk Management
  (Focus and audience on risk is at different levels)
- Fragmented organization structures
- Lack of competent personnel – too close to processes to realize risk
- Inadequate software tools across organization – less visibility

No. 6

CASE STUDY

- Mid-tier upstream company from Kazakhstan
- Compliance based work culture (Tick box mentality)
- Working in silos and fragmented organization
- Training and competency issues

No. 7

HOW DID WE DO IT ?

No. 8

LESSONS LEARNED

- Top down approach on risk is very effective
- Central risk review authority helps visualize the risk across organization
- Enterprise risk and operational risk need to be embedded into one entity
- Training, competency and detailed vocabulary helps for unified risk ranking
- Local language translation (Russian) in this case helped
- Software tool like DNV Synergi Risk helped to raise risk visibility

No. 9

No. 10

Promotion of Process Safety Competency for Operators

- Conference: CCPS Middle East Conference on Process Safety
- Year: 2017
- Time:
  Tuesday, October 10, 2017 - 4:30pm-5:00pm

Authors:
Almaiyad, S., SABIC-IBN SINA

Developing, sustaining, and enhancing the organizationâs process safety competency is one of five elements in the CCPS Risk Based Process Safety pillar of committing to process safety. Organization shall make sure that front line employee (i.e. operators) are competent to handle risk associated with the process.

Ibn Sina has developed process safety competency matrix and embedded it the Job Qualification Program (JQP) of operators for all levels. The competency matrix includes understanding the process risk, consequence of deviation from operating limits, the importance of robust management of change procedure, etc. The JQP uses questioners to assess the level of understanding and fulfilling of competency requirement for each level.

This paper discusses Ibn Sina initiative in development of process safety competency matrix for operators.

ioMosaic
Minimizing risk. Maximizing potential.®
CCPS Middle East Process Safety Conference - 09th -11th October 2017, Bahrain
How to minimize Audit Findings by achieving a Sustainable and Sound Process Safety Culture
Karen Jawich, Safety and Risk Management Consultant
jawich.k.bh@ioMosaic.com
Dr. Gemma Dunjó, Manager of Environmental Services
dunjo.g.nh@iomosaic.com
Elena Prats, CCPSC, Partner
prats.e.nh@ioMosaic.com
© ioMosaic Corporation
Any information contained in this document is copyrighted, proprietary, and confidential in nature belonging exclusively to ioMosaic Corporation.
Any reproduction, circulation, or redistribution is strictly prohibited without explicit written permission of ioMosaic Corporation.

Introduction:
Purpose, Objectives and Motivation

Our main purpose with today's presentation is:

- What?
  - To highlight the key role of a sound process safety culture
- How?
  - By minimizing or avoiding accidental releases of highly hazardous chemicals (HHC)
- Why?
  - To avoid potential catastrophic consequences

April 20, 2010 – Deepwater Horizon (Gulf of Mexico)

Image: Public Domain

What are the specific objectives?

- To **compile and statistically process** audit findings and related data from a sample of Chemical Process Industry (CPI) facilities (ioMosaic's clients)
- To identify the main **common audit findings** and compare them with the results of those analyzed by OSHA (Refinery NEP and Chem NEP)
- To identify **process safety culture benefits** towards minimizing or avoiding these audit findings and therefore, to contributing to an optimized Process Safety management system

Too many major accidents have taken place in industry over the last 40 years

Location	Date	Company	Process	Major Incident	Fatalities (F)/Injuries (I)
Flixborough (UK)	6/1/74	Nypro (UK) Ltd	Production of caprolactam	Explosion in oxidation of cyclohexane process	F: 28 workers I: 36 on-site, 53 off-site
Seveso (Italy)	7/10/76	Industrie Chimiche Meda Societa Azionara (ICMESA)	Batch production of 2,4,5-trichlorophenol (TCP)	Toxic release of TCDD (2,3,7,8-tetrachlorodibenzo-p-dioxin)	F: 0 I: 477 people reported skin injuries (burns & chloracne)
Bhopal (India)	12/3/84	Union Carbide India Ltd	Production of Sevin	Toxic release of methyl isocyanate (MIC)	F: 3,787+ workers and near-by Residents
Piper Alpha (UK)	7/6/88	Occidental Petroleum (Caledonia) Ltd	Offshore oil and gas processing	Oil platform explosion and fire	F: 167 workers
Pasadena (USA)	10/23/89	Phillips 66	Polyethylene production	Polyethylene plant explosion and fire	F: 23 workers I: 130 to 300
Longford (Australia)	9/25/98	Esso Australia Resources Ltd	Gas and crude oil processing	Gas plant explosion and fire	F: 2 workers I: 8
Texas City (USA)	3/23/05	BP	Oil refinery	Isomerisation unit explosion	F: 15 workers I: 180

ioMosaic 6

Process Safety Management is about Accident Prevention

- Major accidents have been driving forces for:
  - Issuing Regulations (Government)
  - Publishing Standards (Industry groups), and
  - Developing Policies (Companies)
- Evolution of Process Safety & Accident/ Loss Prevention Strategies

Standards-based Strategy → Compliance-based Strategy → Continuous Improvement-based Strategy → Risk-based Strategy

Standards-based Strategy	Compliance-based Strategy	Continuous Improvement-based Strategy	Risk-based Strategy
What should I do?	What do I have to do?	How can I improve based on my experience?	How can I better manage risk?

The OSHA Process Safety Management Standard was promulgated in 1992 - 29 CFR 1910.119

- It emphasizes the management of hazards through an established comprehensive **performance based program** that integrates technologies, procedures, and management practices
- If a process uses, stores, manufactures, handles, or moves **highly hazardous chemicals** onsite above the TQ, the process is PSM covered
- Focuses on threats to onsite **employees and contractors**
- Prevents or minimizes the consequences of a **catastrophic release** of toxic, reactive, flammable, or explosive chemicals

The current OSHA PSM 1910.119 standard has 14 elements

1) Employee Participation (EP)
2) Process Safety Information (PSI)
3) Process Hazard Analysis (PHA)
4) Operating Procedures (OP)
5) Training
6) Contractor Safety
7) Pre-Startup Safety Review (PSSR)
8) Mechanical Integrity (MI)
9) Hot Work Program
10) Management of Change (MOC)
11) Incident Investigation
12) Emergency Planning and Response (ER)
13) Compliance Audits
14) Trade Secrets

ioMosaic 9

In response to major accidents, OSHA has implemented several National Emphasis Programs (NEPs)

- 2007 – Petroleum Refinery Process Safety Management NEP
  - In depth audit with static question list
  - **Covered all refineries**
- 2009 – PSM Covered Chemical Facilities NEP
  - Pilot program focused on ammonia and chlorine processes
  - **Regions I, VII, and X**
- 2011 – PSM Covered Chemical Facilities NEP
  - **Expanded to all OSHA Regions**
  - Focused on ammonia refrigeration and all other HHC facilities

ioMosaic 10

ioMosaic's PSM Compliance Audits

Case Study: ioMosaic has conducted many audits over the years. Here we present the audit findings analysis of several Chemical facilities

- A sample of **16 facilities audited**
- Timeframe: **2010 - 2016**
- Type of facilities: **chemical, explosives, refineries…**

Our objective was to evaluate how well each facility complied with the requirements of the OSHA PSM Standard (29 CFR 1910.119)

- The scope of the audit included the **14 PSM elements**, plus an assessment of the applicability of the standard based on the chemicals handled on site
- All 14 OSHA PSM Elements were distributed among the several members of the Audit Team

ioMosaic's audit approach consists of a pre-audit questionnaire, opening, daily debriefing and closeout meetings

- Prior to arriving on-site, a **pre-audit questionnaire** was completed and forwarded to the audit team to help prepare for the audit
- An **opening meeting** was held to communicate to plant personnel the scope and approach for the audit
- **Daily debriefing meetings** were held to communicate preliminary findings and observations made by the audit team
- A **closeout meeting** was held to present all of the regulatory, RAGAGEP and local attention findings

What is an audit Finding, Citation or Violation?

- It is the identification of a part of the PSM program that does **NOT meet regulatory requirements** or industry standards
- Here are finding's categories:
  - **Regulatory** – related to the OSHA PSM Standard
  - **RAGAGEP** – related to Recognized and Generally Accepted Good Engineering Practice (or best industry practice)
  - **Local Attention** – a relatively minor nature that does not represent a chronic PSM issue or a finding related to requirements not specifically listed in the PSM regulation
  - **Compliant** – no evidence of non-compliance was identified

ioMosaic's audit protocol: ioAuditor™, is a component of Process Safety Office™

- ioAuditor™ is a **Windows Excel application** for Safety and Environmental Compliance Audits.
- ioAuditor™ allows recording the results of any type of audit including **PSM and Risk Management Program** (RMP) compliance audits.
- ioAuditor™ has **pre-populated audit protocols** (templates) to simplify documentation of findings and communication to the client

ioMosaic's audit protocol: ioAuditor™

Company:	Confidential	Audit Start Date:	15-Apr-13
Facility:	Confidential	Audit End Date:	18-Apr-13
Unit/Name of Process:	Confidential	Name and Company of Auditors:	ioMosaic Corporation
		Period under Review:	March 11, 2010 - April 18, 2010
Regulatory:	#NAME?	Local Attention:	#NAME?
RAGAGEP:	#NAME?	Compliant:	#NAME?

Item No.	Ref	Element	OSHA 1910.119 Requirements (RMP requirements same except as noted in RMP (ref) below.	Auditor guidance	Emphasis	Findings	Category	Action	Responsible	Due Date
	(n)		Emergency planning and response.							
1.001	(n)	ER	The employer shall establish and implement an emergency action plan for the entire plant in accordance with the provisions of 29 CFR 1910.38(a). In addition, the emergency action plan shall include procedures for handling small releases. Employers covered under this standard may also be subject to the hazardous waste and emergency response provisions contained in 29 CFR 1910.120(a), (p) and (q).	The only new requirement here is the need for a procedure for handling small releases. Verify there is a written emergency action plan that meets the requirements of 29 CFR 1910.38(a). Verify through employee interviews that all employees understand how to protect themselves in an emergency. Verify that places of safe refuge have been identified and safe distances outlined in the plan. Verify that assembly areas are adequately remote from the process. Observe adequacy of windsocks.	NEP	Picked up five emergency plans at different locations in the main office and they were not controlled documents because they did not have blue headers on each page. The reader cannot be sure that these plans contain the most recent update to the plan	Regulatory	Ensure the emergency plan in the current binders is the up-to-date version. Issue updated emergency plans in accordance with controlled document printing requirements.		
1.002	(n)	ER	The employer shall establish and implement an emergency action plan for the entire plant in accordance with the provisions of 29 CFR 1910.38(a). In addition, the emergency action plan shall include procedures for handling small releases. Employers covered under this standard may also be subject to the hazardous waste and emergency response provisions contained in 29 CFR 1910.120(a), (p) and (q).	The only new requirement here is the need for a procedure for handling small releases. Verify there is a written emergency action plan that meets the requirements of 29 CFR 1910.38(a). Verify through employee interviews that all employees understand how to protect themselves in an emergency. Verify that places of safe refuge have been identified and safe distances outlined in the plan. Verify that assembly areas are adequately remote from the process. Observe adequacy of windsocks.	NEP	The Emergency Plan binders contain two documents (one with EP-29 pages, and another with 42 pages). There is no table of contents at the front of the binder or quick tabs that show the organization for easy access to a procedure or section during an emergency.	Local Attention	Determine organization for emergency plan for quick reference during an emergency. Update all emergency plan binders.		
1.003	(n)	ER	The employer shall establish and implement an emergency action plan for the entire plant in accordance with the provisions of 29 CFR 1910.38(a). In addition, the emergency action plan shall include procedures for handling small releases.	The only new requirement here is the need for a procedure for handling small releases. Verify there is a written emergency action plan that meets the requirements of 29 CFR 1910.38(a). Verify through employee interviews that all employees understand how to protect themselves in an emergency. Verify that places	NEP	Recommendation and action item from 2010 audit was not adequately addressed: "The current plan should be revised to include or refer to other procedures regarding the actions to be taken to address small releases of covered process hazardous materials."	Regulatory	Update emergency plan to include response to release of anhydrous ammonia.		

TOC | APP1 | EP1 | PSI1 | PHA1 | OP1 | T1 | C1 | PSSR1 | MI1 | HWP1 | MOC1 | II1 | EPR1 | CA1 | TS1 | Options | Compliant | Local Attention | RAGAGEP | Regulatory | PSM

Audit Findings: Statistical Analysis

Four PSM elements represent more than the 50% of the Regulatory findings

- This chart shows the total % average of Regulatory findings per each of the 14 OSHA PSM elements

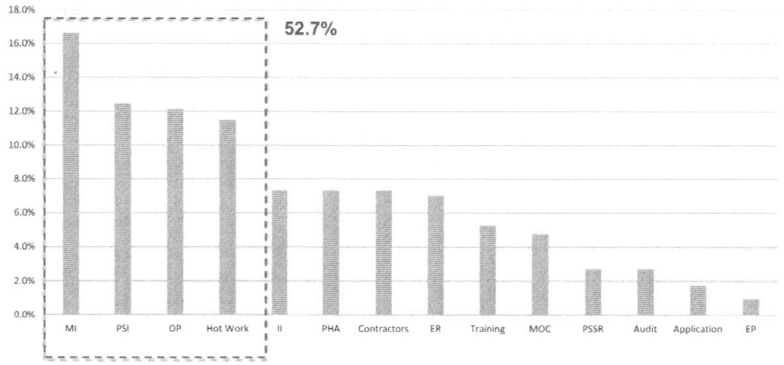

MI: Mechanical Integrity

PSI: Process Safety Information

OP: Operating Procedures

Hot Work

ioMosaic 19

Four PSM elements represent more than the 50% of the RAGAGEP findings

- This chart shows the total % average of RAGAGEP findings per each of the 14 OSHA PSM elements

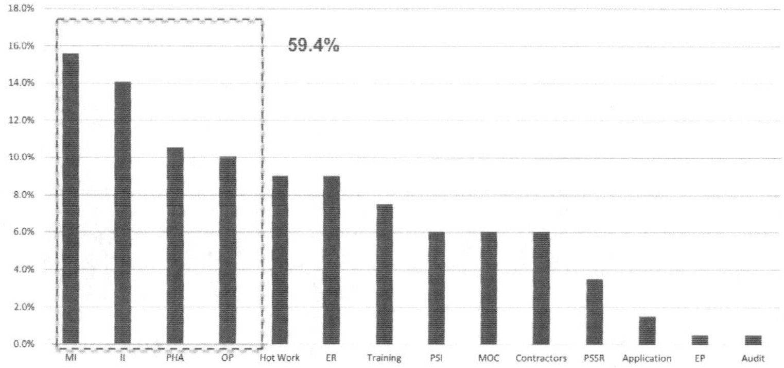

MI: Mechanical Integrity

II: Incident Investigation

PHA: Process Hazard Analysis

OP: Operating Procedures

ioMosaic 20

Four PSM elements represent the 50% of the Local Attention findings

- This chart shows the total % average of Local Attention findings per each of the 14 OSHA PSM element

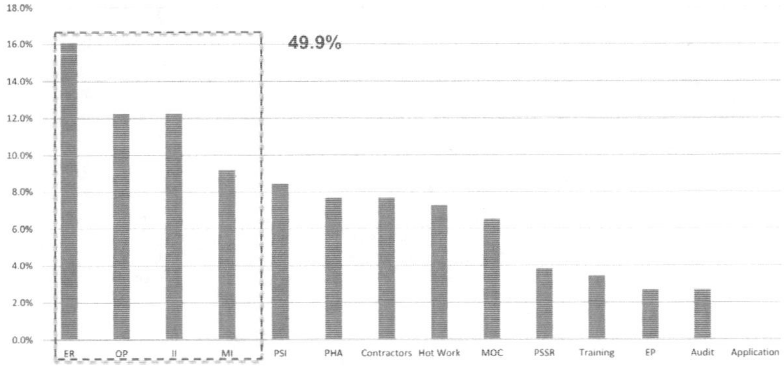

ER: Emergency Planning and Response

OP: Operating Procedures

II: Incident Investigation

MI: Mechanical Integrity

Identification of the main common PSM elements in the 3 categories:

Regulatory:	RAGAGEP:	Local Attention:
MI (16.6%)	MI (15.6%)	MI (9.2%)
OP (12.1%)	OP (10.1%)	OP (12.3%)
II (7.3%)	II (14.1%)	II (12.3%)
PSI (12.5%)	-	PSI (8.4%)
Hot Work (11.5%)	Hot Work (9%)	-
	PHA (10.6%)	ER (16.1%)

MI: Mechanical Integrity

OP: Operating Procedures

II: Incident Investigation

PSI: Process Safety Information

PHA: Process Hazard Analysis

ER: Emergency Preparedness and Response

How are the auditing findings distributed?

› The compilation of findings for all facilities identifies the Regulatory with the main findings, followed by LA and RAGAGEP

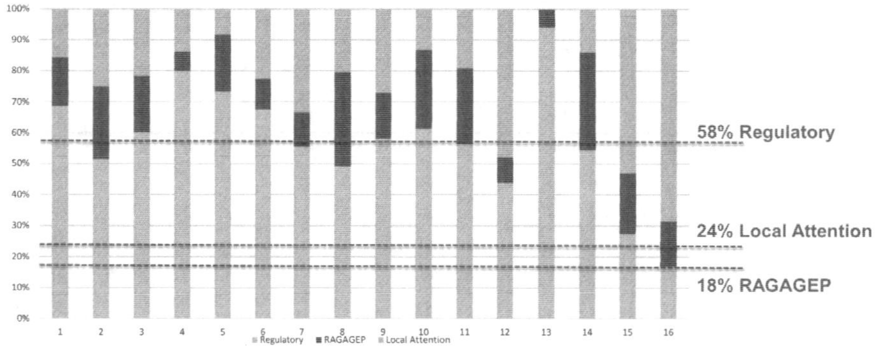

Contribution of each PSM element to the total audit findings

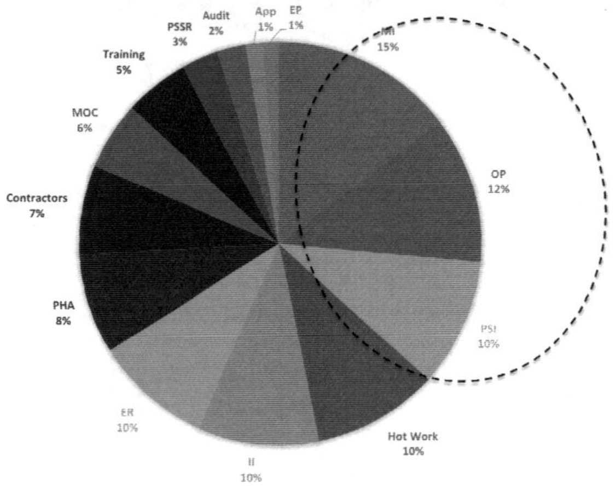

MI: Mechanical Integrity

OP: Operating Procedures

PSI: Process Safety Information

II: Incident Investigation

ER: Emergency Preparedness and Response

PHA: Process Hazard Analysis

Comparison of OSHA Refinery and Chem NEP top PSM elements citations with ioMosaic's findings

%	Refinery NEP	Chem NEP	ioMosaic (Regulatory)
MI	19.5	23.2	16.6
PSI	17.4	20.9	12.5
OP	17.1	14	12.1
Total %	**54**	**58.1**	**41.2**

Description	Refinery NEP	Chem NEP	ioMosaic (Regulatory)
Inspections (facilities)	88	173	16
Citations (findings)	962	1487	648
citations/inspections	10.9	8.6	40.5

Source: Presentation by J.Barab (Deputy Assistant Secretary OSHA, 2012): 'OSHA's Refinery & Chemical National Emphasis Programs. http://www.csb.gov/UserFiles/file/Barab%20(OSHA)%20PowerPoint.pdf

Statistical Analysis Conclusions

- **Key audit findings** is a valued source of information for understanding current weaknesses and lessons learned
- The audit findings are **calls for action**, and should be addressed in a timely manner
- Audits reveal a history of repeat findings indicating chronic problems which can only be effectively achieved addressing the **technical and cultural root causes**
- Mechanical Integrity (MI), Process Safety Information (PSI), Incident Investigation (II) and Operating Procedures (OP), are elements that are highly dependent on the personnel's **attitudes and behaviors**

What do we mean by sound Process Safety Culture?

- *"The combination of group values and behaviors that determine the manner in which process safety is managed. A sound process safety culture refers to* ***attitudes and behaviors*** *that support the goal for safer process operations"*

CCPS, Risk Based Process Safety, 2007

- *In other words:*
  - *All the people in a company wanting to do the 'right thing' the 'right way' 'all the time' (always, even when no one is watching!)*

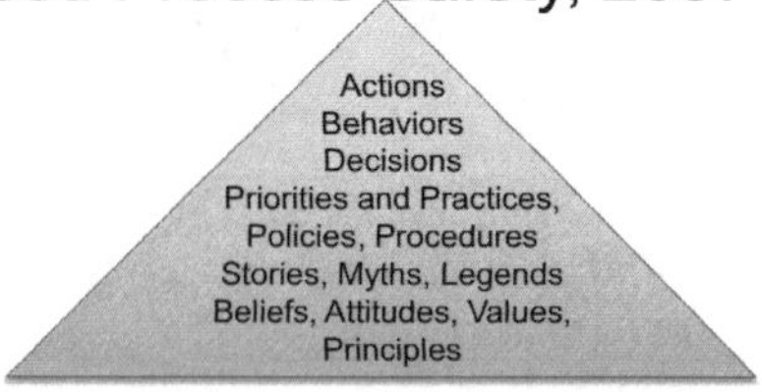

What do we mean by Sustainability?

- *"Sustainable development is development that meets the needs of the present without compromising the ability of future generations to meet their own needs"*

(Brundtland Report, Our Common Future, 1987)

- *Several organizations are promoting sustainable growth via their standards and guidelines*
  - *GRI (Global reporting Initiative Standards)*
  - *ISO 26000 (*International Standard for social responsibility)

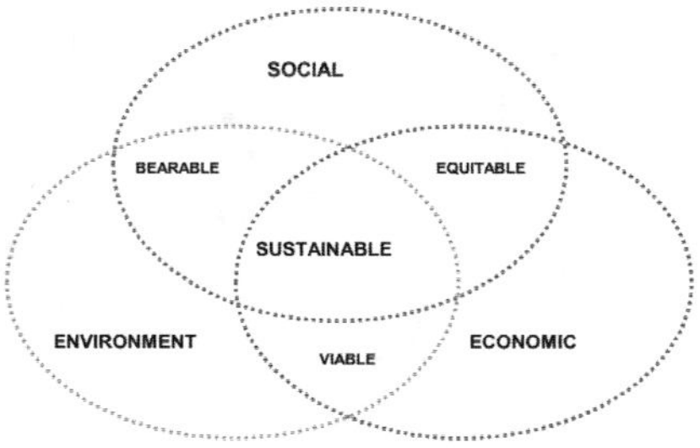

How do we achieve a Sustainable and Sound Process Safety Culture?

- Build a strong Process Safety Management system based on the 14 elements of OSHA PSM 1910.119 standard
- Take into consideration the Environmental, Social, Economic aspects in each process
- Have systems in place to ensure that processes run safely
  - Have no negative environmental impact
  - Allow the protection of the social rights
  - Preserve the economic benefits
- All systems need to be auditable, and the audit findings need to be analyzed and actions need to be taken for a continuous improvement

Example:
Piping Failure in a Crude Unit

Introduction

- When thinning of a piping occurs, ruptures are possible and can lead to the potential release of large quantities of hydrocarbon streams, which might potentially result in catastrophic incidents

Image: Public Domain

What happened?

- Sequence of events:
  - Leaking pipe: an operator notices a small puddle on the ground
  - Operator informs the Head Operator about the leak
    - An Operating Procedure (OP) should address the problem of a leak, by ensuring that the compromised unit is shut down **immediately**
    - Production should never be prioritized over Safety
    - A sound PS Culture ensures that OP are revised and followed. No short cuts!
  - They tried other solutions, but the leak suddenly worsened, hydrocarbon liquid sprayed out of the pipe (flash fire)
  - They decided to shut down the unit but it was too late as a vapor cloud formed and rapidly expanded and the hydrocarbons ignite
    - To act - being knowledgeable in the process (PSI, OP) and in a timely manner is key
  - A plume of black smoke formed, affecting the surrounding community

Observations based on incident investigation reports

- The old carbon steel pipe had thinned to the point of failure from: **sulfidation corrosion**
  - Having accurate Process Safety Information (PSI) is key to Process Safety
  - A sound PS Culture ensures that the operators understand the PSI and act accordingly (e.g. Safety Data Sheets (SDS) give information about corrosion,…)
  - Accurate PSI should be a priority in every company. Companies with a sound PS Culture will establish that P&IDs must be updated every 6 months (Best Practice)
- The same incident had happened at a sister facility a few years earlier
  - Incident Investigation (II) and the resulted lessons learned should help us preventing similar incidents

Observations based on incident investigation reports

- Due to this incident, a Mechanical Integrity (MI) inspection of the piping was performed
  - The result of the MI inspection was severe thinning of the piping
  - The recommendation was replacement. This should have been a priority and never occurred, showing a poor Process Safety Culture
- A few years later, inspectors examined some piping and found significant thinning
  - Each and every segment of a piping should be inspected (MI)
  - If we have identified corrosion, whenever possible the material should be replaced with an inherently safer corrosion-resistant alloy (MI and PSI)
  - If not possible, then piping should be inspected and replaced more often (MI)

General observations

- Overdue inspections should not happen
- Recommendations from inspection should be implemented and in a timely manner
- We should be able to identify the need for inspection and perform it as often as requested (regardless of turnarounds)
- With every Management Of Change (MOC) (e.g. replacement of piping), an update/change of the OP and PSI (e.g. P&IDs, Relief sizing, SDS,...) must be done and communicated accordingly
- Production should never be prioritized over safety

Conclusions

- A safety management program will only be as effective as the underlying safety culture permits
- Change Attitude, safety attitude must be present at all times
- Build Safety Culture Competency
  - If an organization has a weak safety culture, conflicting priorities many arise

- Priority:
  - SAFETY FIRST
  - ZERO TOLERANCE

ioMosaic 37

Questions?

About ioMosaic Corporation

Through innovation and dedication to continual improvement, ioMosaic has become a leading provider of integrated process safety and risk management solutions. ioMosaic has expertise in a wide variety of areas, including pressure relief systems design, process safety management, expert litigation support, laboratory services, training, and software development.

ioMosaic is an integrated process safety and risk management consulting firm focused on helping you manage and reduce episodic risk. Because when safety, efficiency, and compliance are improved, you can sleep better at night. Our over 300 years of industry expertise allow us the flexibility, resources and capabilities to determine what you need to reduce and manage episodic risk, maintain compliance and prevent injuries and catastrophic incidents.

Our mission is to help you protect your people, plant, stakeholder value, and our planet.

For more information on ioMosaic, please visit: **www.ioMosaic.com**

ioMosaic

Minimizing risk. Maximizing potential.®

2017 CCPS Middle East Process Safety Conference, Bahrain

An Overview of Quantitative Risk Assessment Methodology, and Worldwide Risk Tolerability Criteria

Authors:
Jordi Dunjo, Ph.D.
Marcel Amoros
Neil Prophet - Presenter

Contents

- Introduction to Quantitative Risk Assessment
- Conducting a Quantitative Risk Assessment
- Quantitative Risk Assessment Results and Beyond
- Risk Tolerability Criteria
- Conclusions

The concept of Risk has existed for centuries. Quantitative Risk Analysis, as a method, has existed for decades.

- Quantitative Risk Analysis Timeline
- Definition of Risk
  - 1662: Port Royal Logic "Risk should be proportional to both likelihood and consequence"
- Probabilistic Risk Analysis
  - 1967: NASA – introduced methodology following Apollo I accident
  - 1975: Nuclear Regulatory Commission – "Reactor Safety Study"
  - 1981: Nuclear Regulatory Commission – "Fault Tree Handbook"
  - 1983: Nuclear Regulatory Commission – "PRA Procedures Guide"
  - 1985: Nuclear Regulatory Commission – "Probabilistic Safety Analysis Procedures Guide"

The concept of Risk has existed for centuries. Quantitative Risk Analysis, as a method, has existed for decades.

- Chemical Process Quantitative Risk Analysis:
  - Methodology developed throughout 1970s / 1980s
  - Originally based on NUREG guidelines
  - Major chemical industry accidents acted as drivers:
    - Flixborough (1974)
    - Seveso (1977)
    - Mexico City (1984)
    - Bhopal (1984)
    - Piper Alpha (1988)

ioMosaic 5

Quantitative Risk Analysis is typically conducted by following well-defined steps.

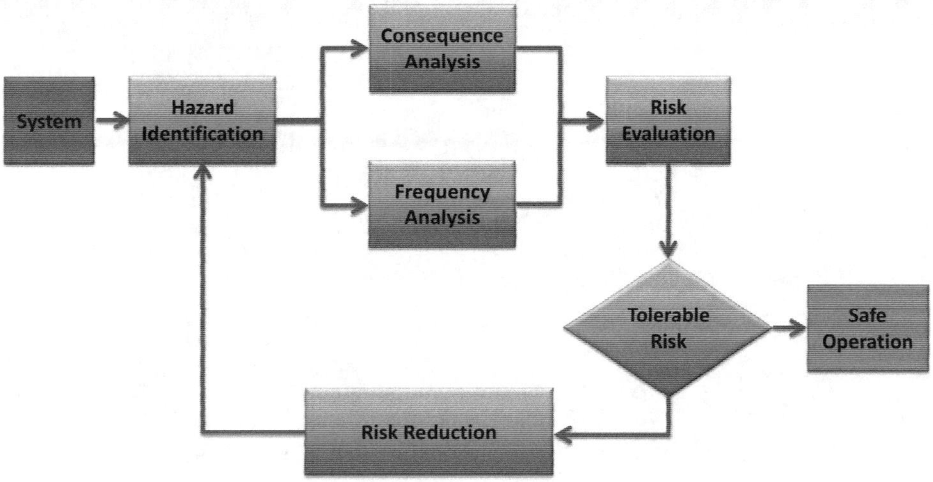

ioMosaic 6

QRA results are typically reported in two forms: Individual Risk and Societal Risk.

- Individual Risk Definition - is the risk of some specified event or agent harming a statistical (or hypothetical) person assumed to have representative characteristics." - HSE, 1995
- Typically assumes recipient is outside 24 hours per day
- Assumes no protective action is taken
- Does not take into account actual population present
- Results typically given as probability of fatality (or dangerous dose) per year
- Useful in facility siting, and land-use planning

ioMosaic 7

QRA results are typically reported in two forms: Individual Risk and Societal Risk.

- Individual Risk results are typically shown in the form of a risk contour

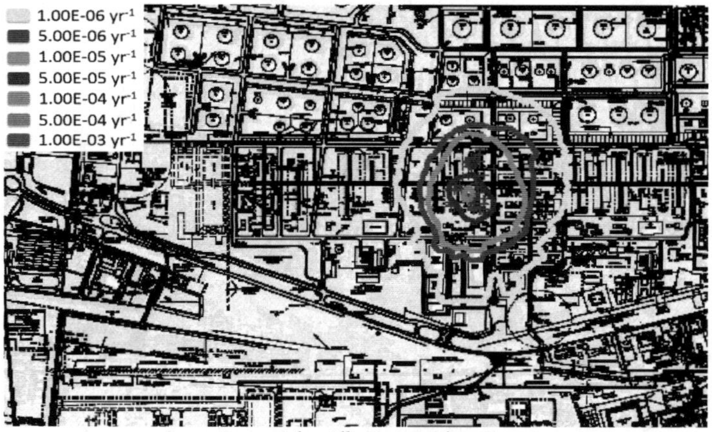

Source: Process Safety Office™ - SuperChems™

ioMosaic 8

QRA results are typically reported in two forms: Individual Risk and Societal Risk.

- Societal Risk Definition - The risk of widespread or large scale detriment from the realization of a defined hazard." - HSE, 1995
- Takes into account actual population present
- Shows frequency (F) of accidents involving N or more fatalities (F-N curve)
- Can be presented as a single number: Average Rate of Death (ROD), or Potential Loss of Life (PLL)
- Technique can be modified to show financial risk rather than fatality risk

QRA results are typically reported in two forms: Individual Risk and Societal Risk.

- Societal Risk results are typically shown in the form of a F-N curve:

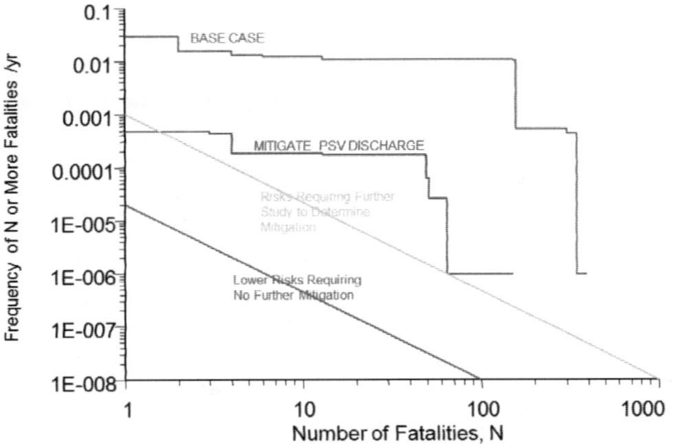

Source: Process Safety Office™ - SuperChems™

A QRA project typically follows these steps:

- Data Gathering
- Hazard Scenario Identification
- Frequency Estimation
- Consequence Modeling
- Risk Calculations
- Compare with Risk Criteria
- Risk Reduction Recommendations

A wide variety of data is required for a QRA study.

- Piping and Instrumentation Diagrams
- Process Flow Diagrams
- Heat and Material Balances
- Equipment Lists
- Plot Plans
- Population and Building Details
- Ignition Source Details
- Meteorological Data
- PHA studies

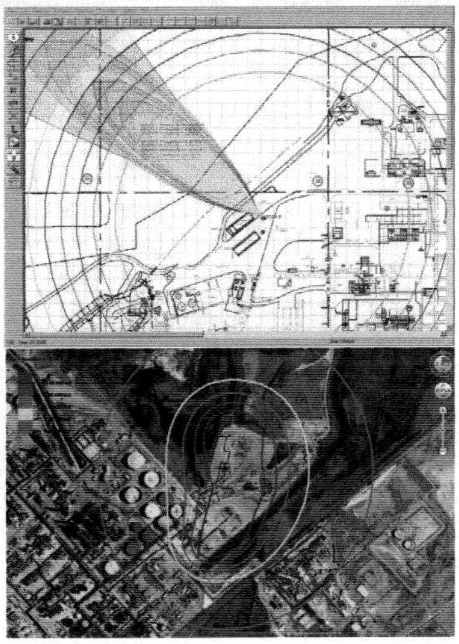

Hazard Scenario Identification considers the potential for "Generic" and "Non-Generic" loss of containment scenarios.

- Generic Scenarios
  - Equipment Failures
  - Piping Failures
- Non-Generic Scenarios
  - Process-specific failures
  - More 'unusual' failures
  - PHA studies can be a good reference for these

Each Loss of Containment (LOC) Scenario may result in multiple outcomes. This video provides a good example.

Each Loss of Containment (LOC) Scenario may result in multiple outcomes.

- Typical QRA hazard outcomes modeled are:
  - Fire – Jet Fire, Pool Fire, Flash Fire, Fireball, BLEVE
  - Overpressure – Vapor Cloud Explosion, Vessel Burst
  - Toxicity – Toxic Effects, Asphyxiation

Each Loss of Containment (LOC) Scenario may result in multiple outcomes. Event tree analysis can be used to determine the probability of each outcome.

› Example Event Tree for Flammable and Toxic Liquid Release:

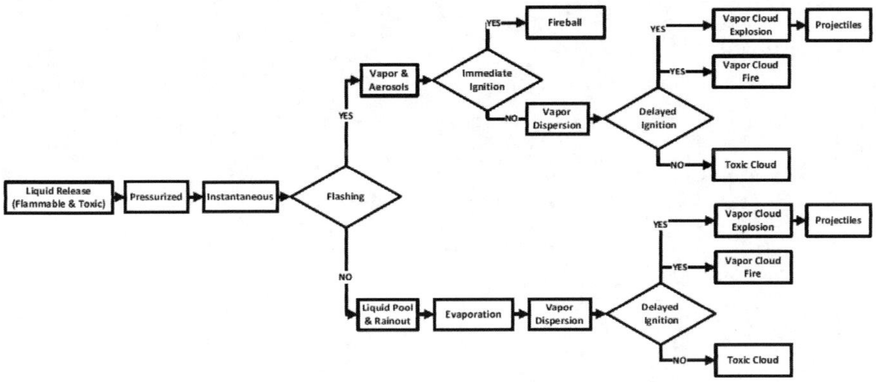

Consequence results can then be evaluated for all identified scenarios.

Vapor Dispersion

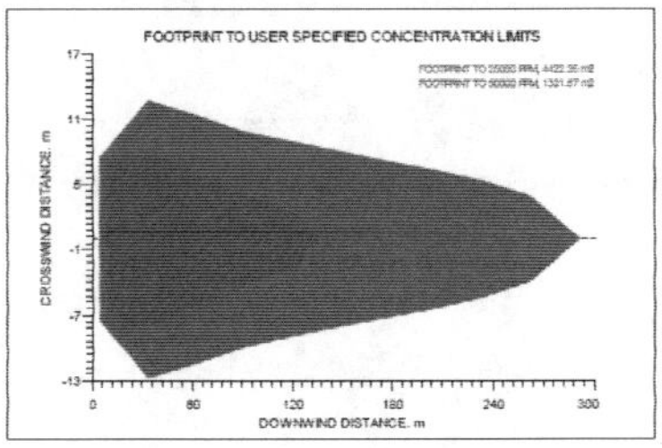

Vapor Cloud Explosion

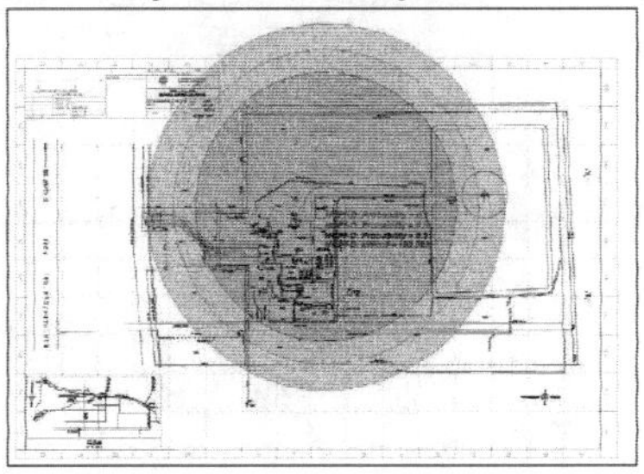

Fireball / BLEVE

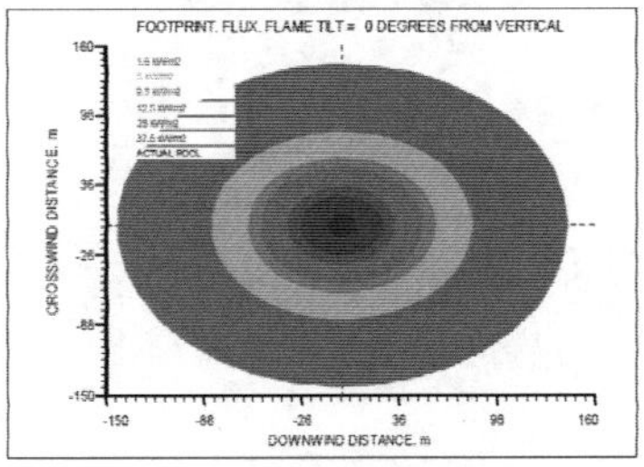

Source: Process Safety Office™ - SuperChems™

Individual risk contours are a key output from a QRA.

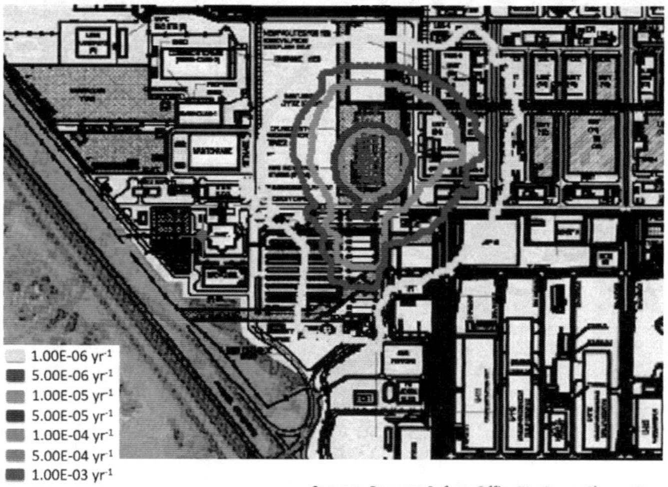

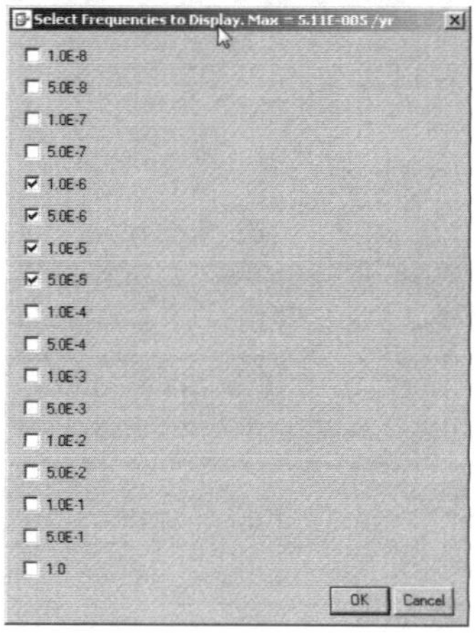

Source: Process Safety Office™ - SuperChems™

Risk transects are also a useful output from a QRA.

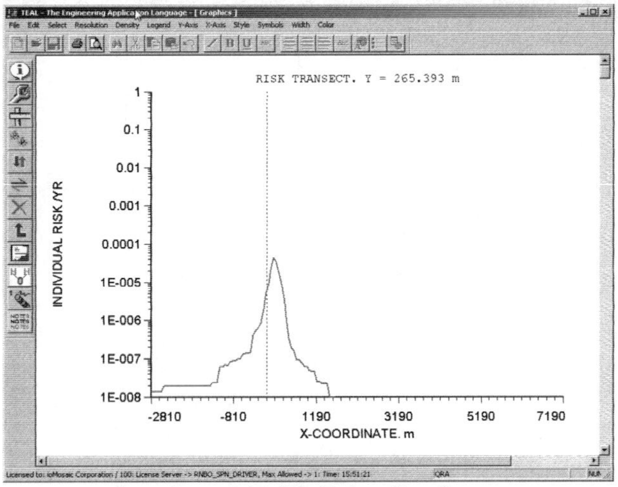

Source: Process Safety Office™ - SuperChems™

ioMosaic 21

Societal risk is also an important output from a QRA.

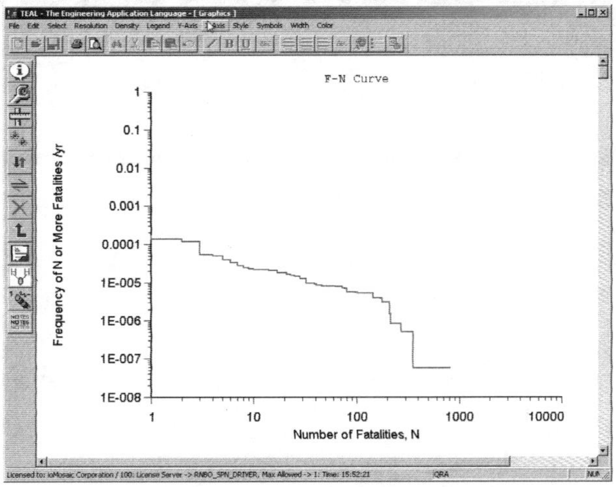

Source: Process Safety Office™ - SuperChems™

ioMosaic 22

Multiple societal risk results can be presented on the same F-N curve, together with acceptance criteria.

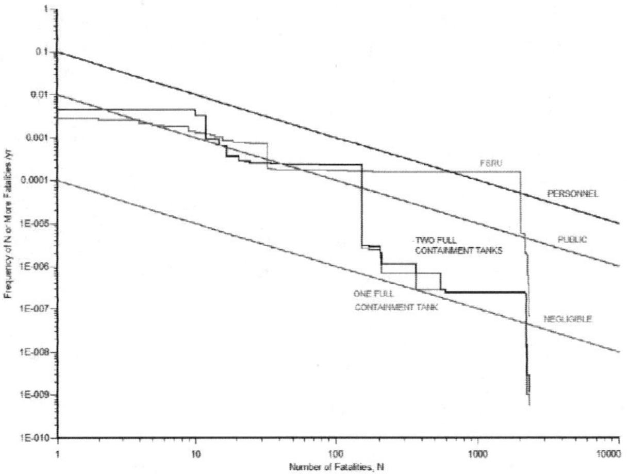

Source: Process Safety Office™ - SuperChems™

Beyond typical QRA results, Overpressure Exceedance Curves can be generated (e.g., for various receptors).

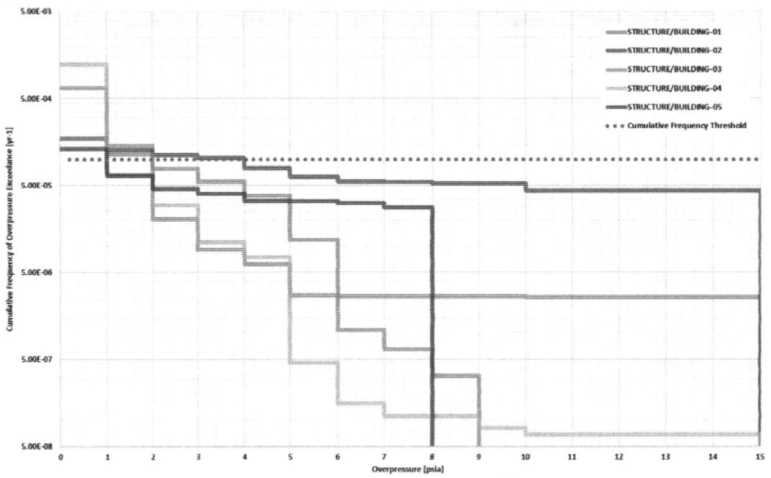

Beyond typical QRA results, a Building Damage Exceedance Curve (BDEC) can then be generated for a given building.

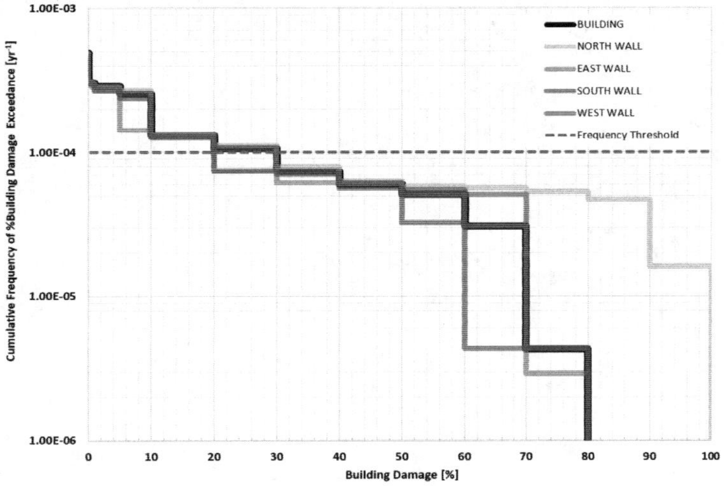

ioMosaic 25

In turn, a F-N curve for a specific building can be created.

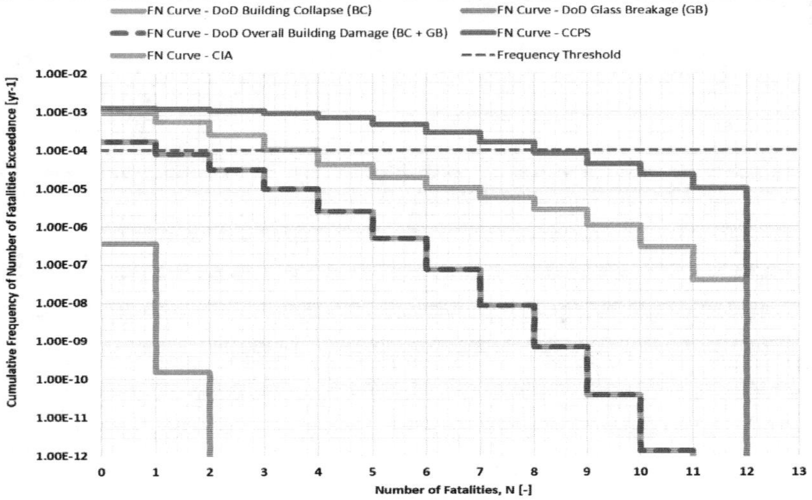

ioMosaic 26

Generally accepted and recognized risk criteria typically incorporates three themes.

- A comprehensive risk management program should address both individual and societal risk
- Risk criteria for the public should be lower, i.e., more conservative, than for the workforce since the workforce risk is considered to be voluntary
- With respect to individual risk, new facilities should be held to a higher level of risk performance than existing facilities

What constitutes "acceptable risk"?

- The phrase "acceptable risk" is widely used in Quantitative Risk Assessment (QRA) literature
  - Individuals may "accept" risk of an activity on a voluntary basis if they deem it is low enough and if they derive a benefit from it
  - When a risk from an activity is imposed on an individual on an involuntary basis and there are no perceived benefits to the individual, then no risk would be considered truly "acceptable" no matter how small

ioMosaic

What constitutes "acceptable risk"?

- Factors affecting risk acceptability:
  - Economic benefit
  - Amenities
  - Voluntary or involuntary risk
  - Visible risk
  - Size of potential accidents

ioMosaic 30

Risk Tolerability Regions – ALARP Definition.

- Unacceptable Region
  - Except under extraordinary circumstances, control measures must be undertaken to reduce the risk to a level deemed tolerable irrespective of the cost/benefit
- Tolerable Region
  - The residual risk must be at a level ALARP; i.e., "risk must be averted unless there is a gross disproportion between the costs and benefits of doing so"
- Broadly Acceptable Region
  - Residual risk is generally regarded as insignificant and adequately controlled

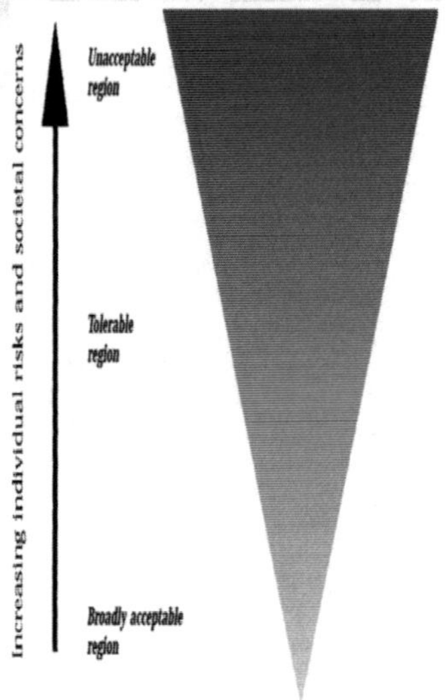

Worldwide Risk Tolerability – Variety of Criteria.

- Individual Risk
  - Risk to the Public – Entities having two-limit system: Upper and Lower Limits
  - Risk to the Public – Entities having single-limit system: Upper Limit
  - Risk to the Workforce – Entities having two-limit system: Upper and Lower Limits
  - Risk to the Workforce – Entities having single-limit system: Upper Limit
- Societal Risk
  - Varying risk tolerability lines in use
  - Different criteria for Public and Workforce

Worldwide Risk Tolerability – Select Examples Are Provided.

- ioMosaic has performed a review of the state-of-the-art of risk criteria from worldwide entities and applications
- Some select examples are provided for
  - USA
  - United Kingdom
  - France
  - Canada

Risk Criteria is in use within the USA in certain applications.

- No federal risk based criteria
- California
  - LACFD defines criteria for risks associated with their RMP program
  - County of Santa Barbara
    - Public safety thresholds based on F-N curves
    - Zoning based on risk for land use planning (red, amber, green zones)

Risk Criteria is in use within the USA in certain applications.

- National Fire Protection Agency – 59A – Standard for the Production, Storage and handling of Liquefied Natural Gas (LNG)

Criterion Annual Frequency	Remarks
Zone 1 Risk ≤ 1.00E-05	**Not Permitted:** residential, office, and retail **Permitted:** occasionally occupied development (e.g., pump houses, transformer stations)
Zone 2 1.00E-06 ≤ Risk ≤ 1.00E-05	**Not Permitted:** shopping centers, large-scale retail outlets, restaurants, etc. **Permitted:** work places, retail and ancillary services, residences in areas of 28 to 90 people/hectare density
Zone 3 3.00E-07 ≤ Risk ≤ 1.00E-06	**Not Permitted:** churches, schools, hospitals, major public assembly areas, and other sensitive establishments **Permitted:** all other structures and activities

Risk Criteria is well established in the United Kingdom.

- HSE Planning Advice for Developments Near Hazardous Installations

Sensitivity	Description and Examples	Criteria
Level 1	Based on normal working population – parking areas, warehouses, non-retail, less than 100 occupants, minor transportation links.	Ok in all zones < 1.00E-05
Level 2	Based on the general public – at home and involved in normal activities – residential units less than 40 per hectare, hotels, motels up to 100 beds, major transport links, retail less than 5000 m2, gatherings of less than 100 people.	Ok in middle and outer zones only < 1.00E-06
Level 3	Based on vulnerable members of the public (children, those with mobility difficulties or those unable to recognize physical danger) – more than 100 beds, more than 40 units per hectare, more than 100 people outdoors, hospitals 24 hr care < 0.25 hectare prisons.	Ok in outer zone only <3.00E-07
Level 4	Large examples of Level 3 and large outdoor examples of Level 2 – theme parks, stadiums, open air areas with more than 1000 people, hospitals > 0.25 hectare, daycare larger than 1.4 hectare.	Not ok in any zone

Risk Criteria is well established in the United Kingdom.

› HSE Planning Advice for Developments Near Hazardous Installations

Frequency	Zone	Description
< 1.00E-05	Inner Zone	Receiving a "dangerous dose" or worse.
< 1.00E-06	Middle Zone	Receiving a "dangerous dose" or worse.
< 3.00E-07	Outer Zone	Receiving a "dangerous dose" or worse. This criterion is appropriate for highly vulnerable or very large public facilities.

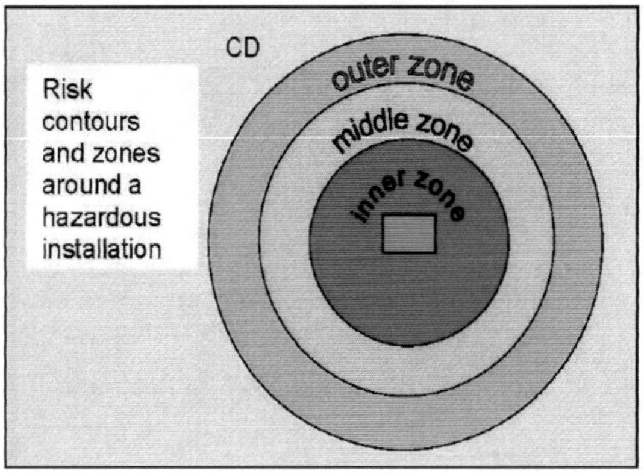

ioMosaic 37

Risk Criteria is used in France.

› Code de l'Environnement

Type of Risks and Facilities	Type of Accident Scenario
Risks linked to liquefied combustible gas facilities (fixed, semi-mobile or mobile)	**Scenario A:** BLEVE (Boiling Liquid Expanding Vapor Explosion) **Scenario B:** VCE (Vapor Cloud Explosion)
Risks linked to vessels containing liquefied or non-liquefied toxic gases where the containment is not designed to resist external damage or internal reactions of products	**Scenario C:** Total instantaneous loss of containment
Risks linked to vessels containing toxic gases where the containment is designed to resist external damage or internal reactions of products	**Scenario D:** Instantaneous rupture of the largest pipeline leading to the highest mass flow
Risks linked to large vessels containing flammable liquids	**Scenario E:** •Fire in the largest tank •Explosion of the gas phase for fixed roof tanks •Fireball and projection of burning product due to boil-over
Risks linked to use or storage of explosives	**Scenario F:** Explosion of the largest mass of explosive present or explosion due to a reaction

ioMosaic 38

Risk Criteria is used in Canada.

- Risk-based Land Use Planning Guidelines - Major Industrial Accidents Council of Canada

Frequency Level	Type of Zone	Allowed Land Uses
>1.00E-04	Buffer zone	None
>1.00E-05	Municipality transition zone	Manufacturing, open spaces, golf courses
>1.00E-06	Municipality transition zone	Commercial, low density residential
<1.00E-06		All other uses

Risk Tolerability Regions Establish Upper and Lower Limits.

- Clarifications
  - Upper Limit: High Risk Region; i.e., if risk level greater than the upper limit, then the risks are Unacceptable
  - Lower Limit: Negligible Risk Region; i.e., if risk level lower than lower limit, then the risks are Broadly Acceptable

Individual Risk (Public) – Entities having two-limit system.

Upper Limit [Fatality·yr⁻¹]	Lower Limit [Fatality·yr⁻¹]	Entities/Applications
1.00E-04	1.00E-05	State of Sao Paulo, Brazil/Pipelines
1.00E-04	1.00E-05	State of Rio Grande do Sul, Brazil/Pipelines
1.00E-04	1.00E-06	UK HSE/Fixed facilities and dangerous goods transport
1.00E-04	1.00E-06	International Maritime Organization (IMO)/Existing ships
1.00E-05	1.00E-07	State of Victoria, Australia
1.00E-05	1.00E-6	State of Sao Paulo, Brazil/Fixed installations
1.00E-05	1.00E-06	State of Rio Grande do Sul, Brazil/Fixed installations
1.00E-05	1.00E-06	Hungary
1.00E-05	1.00E-06	International Maritime Organization (IMO)/New ships

ioMosaic 41

Individual Risk (Public) – Entities having single-limit system (1).

Upper Limit [Fatality·yr⁻¹]	Entities/Applications	Comments
5.00E-05	Australia, State of Queensland	Specifies that 5.00E-05 fatality/year risk contour must not extend beyond site boundary for new facilities. For existing facilities, risk reduction is to be "encouraged" if 5.00E-5 fatality/year risk contour extends beyond site boundary.
5.00E-05	Singapore	Specifies that the 5.00E-05 fatality/year risk contour may only extend into industrial development zones.
1.00E-05	Hong Kong	New installations. Existing installations exceeding this value should seek risk reductions
1.00E-05	Netherlands	Applies to vulnerable objects. Existing situations. Interim value, existing situations must meet value for new situations (1.00E-06 fatality/year, see below) by 2010
1.00E-05	Canada, Major Industrial Accidents Council of Canada (MIACC)	Uses this value for low density residential, and lower value (1.00E-06 fatality/year) for high density residential.
1.00E-05	Czech Republic	Limit for existing installations. Risk reduction must be carried out for facilities above this limit.
1.00E-05	State of Rio de Janeiro, Brazil/Fixed installations and pipelines	For existing facilities
5.00E-06	Singapore	Specifies that the 5.00E-06 fatality/year risk contour may only extend into industrial and commercial development zones.
1.00E-06	Netherlands	Applies to: (1) new permits for fixed installations, (2) new land use plans, and (3) transport of dangerous goods, including transport by pipelines

ioMosaic 42

Individual Risk (Public) – Entities having single-limit system (2).

Upper Limit [Fatality·yr^{-1}]	Entities/Applications	Comments
1.00E-06	State of Western Australia, Australia	New installations. Higher limits are established for industrial and non-industrial developments. Lower limits (5.00E-07 fatality/year) are established for "sensitive" exposures. Existing installations are to seek risk reductions to meet requirements for new installations.
1.00E-06	State of New South Wales, Australia	Higher limits are established for industrial and nonindustrial developments. Lower limits (5.00E-07 fatality/year) are established for "sensitive" exposures.
1.00E-06	State of Queensland, Australia	Higher limits are established for industrial and nonindustrial developments. Lower limits (5.00E-07 fatality/year) are established for "sensitive" exposures.
1.00E-06	State of Rio de Janeiro, Brazil/Fixed installations and pipelines	For new facilities
1.00E-06	Czech Republic	For new facilities
1.00E-06	US DOD/Explosives handling activities	No comments
1.00E-06	California, Santa Barbara County	Used as a screening value. Risk in excess of this value requires a risk assessment examining the societal risk from the facility
1.00E-06	Singapore	Specifies that the 1.00E-06 fatality/year risk contour may only extend into industrial, commercial, and park developments zones.
1.00E-06	Malaysia	Malaysia: higher risk criteria established for those on industrial facilities, exposed from outside source.
1.00E-07	US NRC/Risk of "prompt" fatalities from nuclear power accidents	Calculated from the criterion that risks should not exceed 0.1% of prompt fatality risks from all other accidental sources, assuming an accidental fatality rate of 1.00E-04 fatality/year

Individual Risk (Workforce) – Entities having two-limit system.

Upper Limit [Fatality·yr^{-1}]	Lower Limit [Fatality·yr^{-1}]	Entities/Applications	Comments
1.00E-03	1.00E-06	UK HSE	No comments
1.00E-03	1.00E-06	International Maritime Organization (IMO)/Existing ships	**Upper Limit:** for crew members on existing ships **Lower Limit:** for crew members on new or existing ships
1.00E-03	1.00E-04	State of Western Australia, Australia	**Upper Limit:** proposed for existing facilities. Where an existing facility exceeds 1.00E-03 fatality/year, a risk reduction program with an agreed time frame must be implemented to achieve 1.00E-03 fatality/year **Lower Limit:** proposed for new and existing facilities
5.00E-04	1.00E-04	State of Western Australia, Australia	**Upper Limit:** proposed for new facilities **Lower Limit:** proposed for new and existing facilities
1.00E-04	1.00E-06	International Maritime Organization (IMO)/New ships	**Upper Limit:** for crew members on existing ships **Lower Limit:** for crew members on new or existing ships

Individual Risk (Workforce) – Entities having single-limit system.

Upper Limit [Fatality·yr^{-1}]	Entities/Applications	Comments
1.00E-04	US DOD/Explosives handling activities	No comments

Societal Risk – Graphical Summary of Worldwide Criteria.

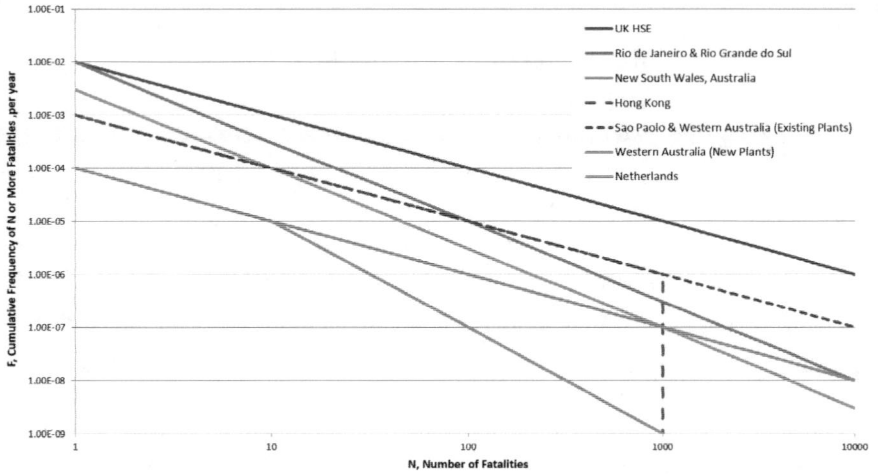

Conclusions

- The risks posed by hazardous industrial facilities are controlled by safety legislation: it is required that all hazards are identified, the risks arising are assessed and controls put in place to minimize the likelihood of an accident.
- Explicit, well-defined risk criteria, defined as part of risk assessment activities, add to the clarity of the decision-making process.
- Communication of risk information can be difficult. Accepted risk criteria should make such communications easier and more effective. Visualization of these risks aids understanding.
- Make sure the public understands the risks being communicated.

Conclusions

الوطن

مشروع «الوقود البيئي».. قنبلة موقوتة

SPARKLE TOWERS

اقتصاد

مشروع «الوقود البيئي» قنبلة موقوتة تهدد العاملين وسكان المناطق القريبة

Conclusions

- Society increasingly expects industry to be cognizant of, and responsibly manage, the risks of its operations.
- As the ability to manage risks increase, the public will expect industry to do better. Consequently, risk criteria is subjected to periodic reconsideration.
- A comprehensive risk management program should address both individual and societal risk.
- Risk criteria for the public should be lower, i.e., more conservative, than for the workforce since the workforce risk is considered to be voluntary.
- With respect to individual risk, new facilities should be held to a higher level of risk performance than existing facilities.

References

- Health and Safety Executive (HSE). Risk Criteria for Land-Use Planning in the Vicinity of Major Industrial Hazards. HMSO, 1989
- AIChE/CCPS, Guidelines for Developing Quantitative Safety Risk Criteria; New York; American Institute of Chemical Engineers / Center for Chemical Process Safety, 2009
- CCPS Guidelines for Evaluating Process Plant Buildings for External Explosions and Fires, AIChE/CCPS, 1996
- API-581, Risk Based Inspection Base Resource Document, 2nd Edition, 2008
- Health and Safety Executive, HSE. Failure Rate and Event Data for use within Risk Assessments, June 2012

About ioMosaic Corporation

Through innovation and dedication to continual improvement, ioMosaic has become a leading provider of integrated process safety and risk management solutions. ioMosaic has expertise in a wide variety of areas, including pressure relief systems design, process safety management, expert litigation support, laboratory services, training, and software development.

ioMosaic is an integrated process safety and risk management consulting firm focused on helping you manage and reduce episodic risk. Because when safety, efficiency, and compliance are improved, you can sleep better at night. Our over 300 years of industry expertise allow us the flexibility, resources and capabilities to determine what you need to reduce and manage episodic risk, maintain compliance and prevent injuries and catastrophic incidents.

Our mission is to help you protect your people, plant, stakeholder value, and our planet.

For more information on ioMosaic, please visit: **www.ioMosaic.com**

ioMosaic 53

Hazard Identification and Risk Assessment during Capital Projects

By: Adel Dakheel
William Bridges
Revonda Tew

For: MEPSC, October 10, 2017, BAHRAIM

Process Improvement Institute, Inc.
adakheel@piii.com
www.piii.com

1

Presenter: Adel Dakheel
Process Improvement Institute (PII)

- BS in Biological Systems Engineering
- 16+ years experience in the chemical/petrochemical industry includes Operations, Project Management, Corporate and Consulting
- Experience in SABIC (Jubail Petrochemical Co., Corporate EHSS, Saudi Kayan Petrochemical Co., and Engineering & Project Management). Key player in initial roll-out of SABIC Process Safety Management System (prior to SHEMS)
- 20+ PSM Audits, many Fire Safety Audits, PSM implementation plans (SABIC and PII)
- Led 20+ Unit scale PHAs, 30+ MOC PHAs; scribed on other PHAs (at PII); Participated on numerous Investigations/RCAs (at SABIC and PII)
- Lead Instructor for PSM, RCA, PSM Auditing, PHA Leadership, and other PSM systems

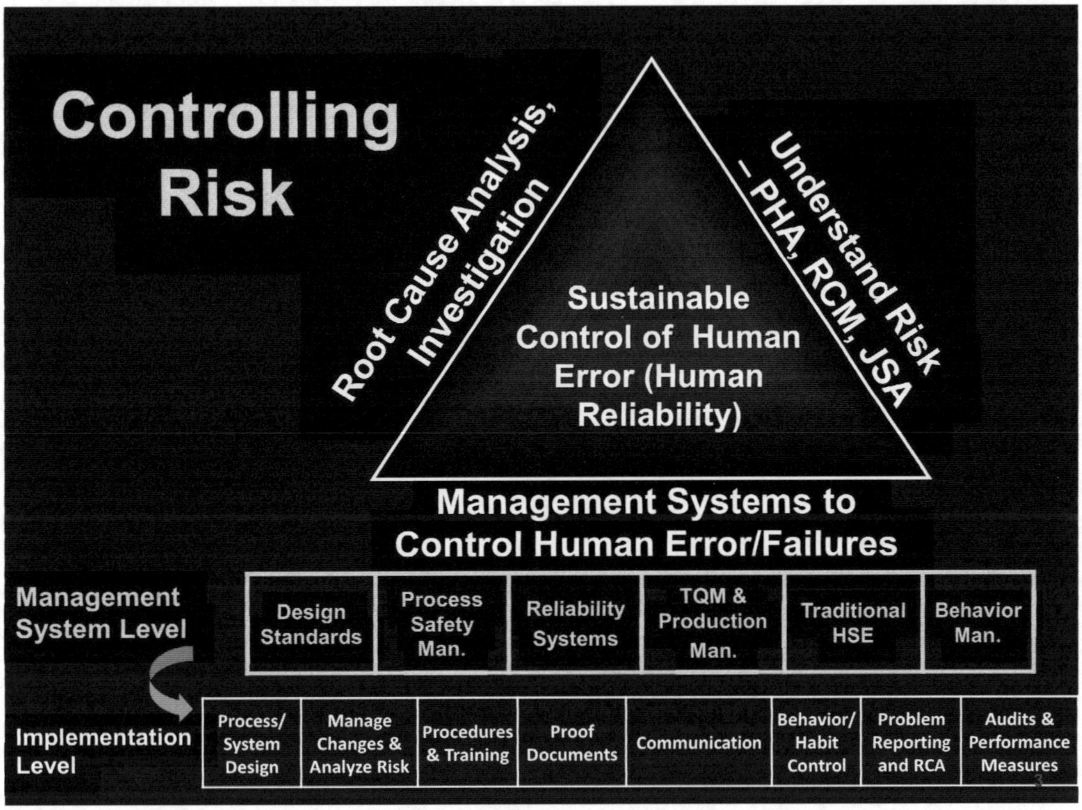

History of Risk Assessment & PSM Development during Projects

- Began (in earnest) in mid-1970s in the chemical industry
- Risk assessments & developing PSM during projects widely accepted by mid-1980s and became an element of CCPS's PSM standard

4

Comparison of PSM Systems

OSHA 29 CFR 1910.119 and EPA 40 CFR 68	AIChE CENTER FOR CHEMICAL PROCESS SAFETY	ACC RESPONSIBLE CARE PROCESS SAEFTY CODE®
Management System Employee Participation† Process Safety Information Process Hazard Analysis Operating Procedures Training Contractors† Pre-startup Safety Review Mechanical Integrity Hot Work Permit† Management of Change Incident Investigation Emergency Planning and Response Compliance Audits Trade Secrets	Accountability Process Knowledge and [illegible] Management of Change Process Equipment Integrity Incident Investigation Training and Performance Human Factors Standards, Codes and Laws Audits and Corrective Actions Enhancement of Process Safety Knowledge	*Management Leadership* Commitment/Accountability Performance Measurement Incident Investigation Information Sharing CAER Integration *Technology* Design Documentation Process Hazards Information Process Hazards Analysis Management of Change *Facilities* Siting Codes and Standards Safety Reviews Maintenance and Inspection Multiple Safeguards Emergency Management *Personnel* Job Skills Safe Work Practices Initial Training Employee Proficiency Fitness for Duty Contractors

Project reviews and design procedures

Example from Process Improvement Institute, Inc., PSM Course, 2004

5

History of Risk Assessment & PSM Development during Projects (cont.)

- **Many excellent papers have been presented on the topic within AIChE/CCPS (Kelly, Broadribb, McGrath, etc.)**
- **Some major companies are still weak**

6

Purpose of this Paper

- Summarize the enduring aspects of managing risk during a project
- Highlight some new ways to create and deliver PSM from a project
- Highlight pitfalls to avoid and current best practices

7

General Concept of Project Risk Management

- Evaluate the risk of a design (for both new or old technology) and ensure there are proper safeguards before startup
- Manage risk from phase-to-phase of a project
- Ensure input of owners/operators
- Ensure project delivers what is necessary to help owners/operators manage risk long-term

8

Types of Projects – Size

Project Size	Example Project Scope	Example Project Length/Duration (concept→ commissioning)	Number of Risk Reviews
Major	Major projects handled external to an affiliate/plant, such as expansions and new facilities	12-36 months	4-7
Medium	Works engineered by an affiliate/ plant (installing a new design of knockout pot for a feed to a unit)	6-9 months	2-3
Small	Minor affiliate/plant works (installing piping to bypass a control valve)	1-2 months	1

9

PSM Deliverables from Project –

works best if understood as same management practices required for excellent operability & reliability

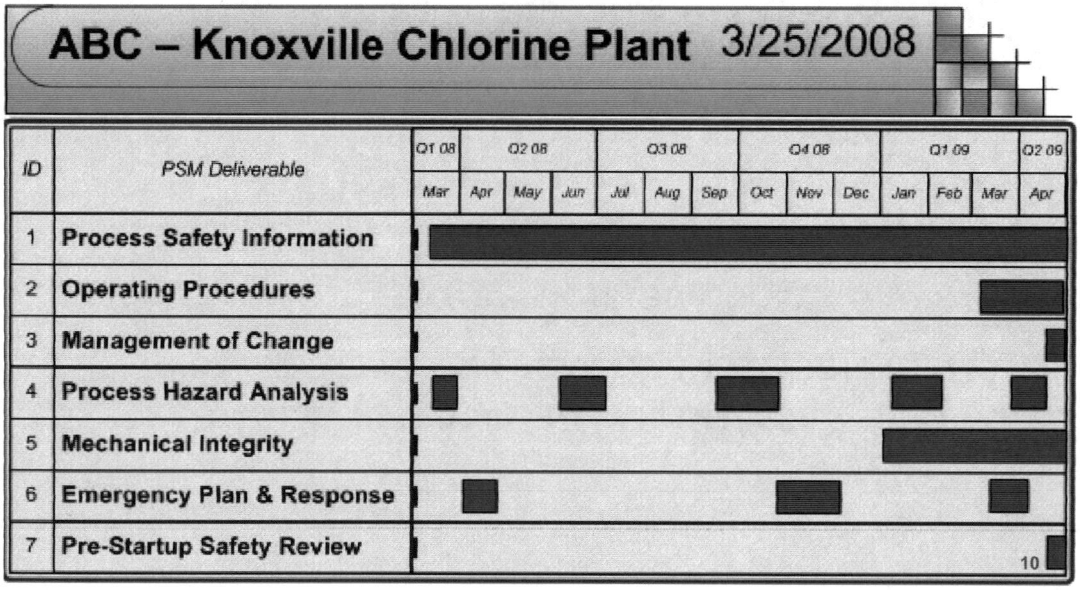

Types of Hazard/Risk Reviews in Life Cycle of a Process (each type uses one or more of the HR (PHA) methods)

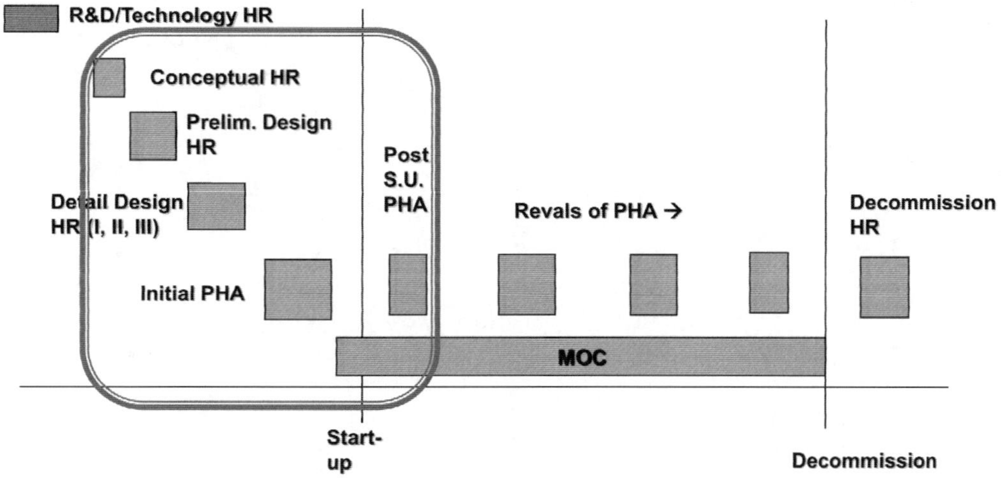

Example from Process Improvement Institute, Inc., PHA Leadership Course, 2003

Time →

11

	Project Responsibility						Plant/Unit Responsibility
	Phase 1 Conceptual Design	**Phase 2** Feasibility & Detailed Specification	**Phase 3** Preliminary Design	**Phase 4** Detailed Design	**Phase 5** Construction	**Phase 6** Pre-Commissioning	**Phase 7** Post-Commissioning
Project Concerns and Deliverables	Fit to business strategic plan Fit to existing operations Review of available technology Inherent safety options Site planning Raw material resourcing options	Detailed feasibility study (availability of technical staff; marketing plan) Detailed technology review and specification development Preliminary plot plan and tie-in plan +/- 40% cost estimate Preliminary schedule & milestones	Preliminary construction & operation plans Initial process flow diagrams (PFDs) Initial material and energy balances Raw material planning Utility planning Candidate vendors for major components Fire protection plan	Process design and generation of Rev 0 P&IDs; continuing to Rev 5 (nominal) in this phase. Revise material & energy balance Technical specs for all components Basics design of process controller Fabrication started and major components ordered	Execute fabrication and installation plan Develop detail plans for tie-ins to OSBL Continue to develop specs for OEM manuals and operating procedures Set up CMMS/ database for ITPM Draft PSM management systems/procedures	Complete fabrication/ installation Complete development and closure of PSSR and other punchlists Commission/ validate equipment (dry, wet, with HHC) Populate CMMS and other PSM/ reliability databases Complete operating and MI procedures (by SMEs); validate	Assist commissioning team Ensure training by vendors/OEMs are completed in the field and proficiency of plant staff is validated Complete performance measure of initial operation (to ensure contract commitments are met) Manage changes Closeout project
Risk Reviews	**Conceptual RR** Strategic plans Inherent safety Plot plan review for Facility Siting; consequence modeling for major releases Begin Human Factor consideration		**Preliminary Design RR** What-if analysis of each major unit operation HAZOP/FMEA of selected scenarios LOPA of selected scenarios & review options for inherent safety	**Detailed Design RR** HAZOP/FMEA of most nodes (focusing on continuous mode of operation) LOPA of 1-5% of scenarios Final SIL (if needed) determination	**Final Detailed Design RR** HAZOP/FMEA of changes since previous RR, including rec. resolutions; place special attention to changes in field Begin human factor and facility siting (HF&FS) checklists	**Commissioning RR ("Initial PHA for new unit")** HAZOP/FMEA of changes HAZOP and/or What-if of non-routine operating modes startup, emergency shutdown, etc.) Complete HF&FS checklists	**Post-Startup RR (3 to 6 months after startup)** Close any recommendations that were rated as post-startup issues Review each MOC for its impact on the "Initial Unit PHA" Perform critique of risk review efforts during project.

12

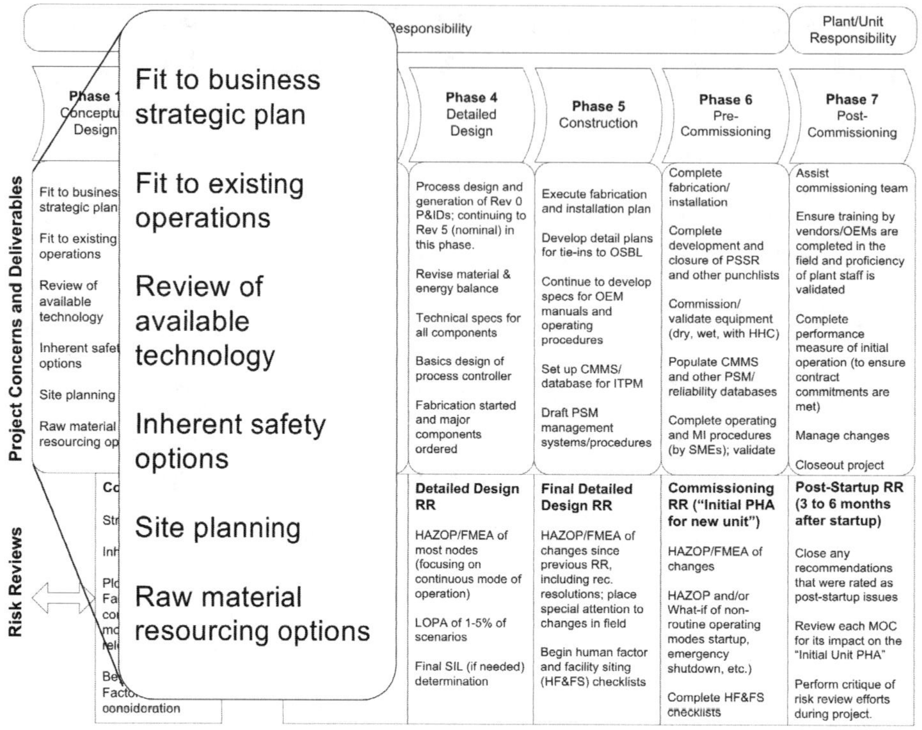

13

Conceptual Design Phase

Key Concern	Ways to Address/Evaluate Key Concern during Design and Risk Reviews
Inherently safe/reliability process selection	▪ Evaluate low waste chemistries ▪ Evaluate low inventory ▪ Evaluate use of non-hazardous solvents and reagents ▪ Evaluate low energy rea…
Plot Location Relative to Other Units	▪ Process opportuniti… ▪ Operating logistics… ▪ Distance from con… ▪ Backup contingen… ▪ Knock-on effects … e modeling is necess… ▪ Begin Facility Siti…
Size of Plot Areas – based on preliminary process designs	▪ Access to utilities… ▪ Cost of real estate… ▪ Elevation conside…
Precise Spacing Between Unit Boundaries – distances between battery limits or between nearby units	▪ Insurance and Ind… ▪ Knock-on effect t… ▪ Business interrupt…
Inner Unit Layout	▪ Ease of Access (be… maintainability ▪ Constructability a… s for construction an… ▪ Equipment decking … aged ladders to take routine p…
Fire Protection Review	▪ Emergency access routes; Fire fighting or toxic rescue capability ▪ Placement of detectors ▪ Strategy to protect neighbors from your releases ▪ Construction of control rooms to protect "stay-behinds"

Example of more detail in the paper

14

PSM Deliverables – Phase 1 & 2: Conceptual Design

- **Process safety information,** including chemical hazards, reactivity, hazards of inadvertent mixing, inventories, applicable codes & standards
- **Baseline info for future PHA**
- **Baseline info for future MI**
- **Begin inherently safer consideration**
- **Begin leadership**
- **Begin employee participation**

15

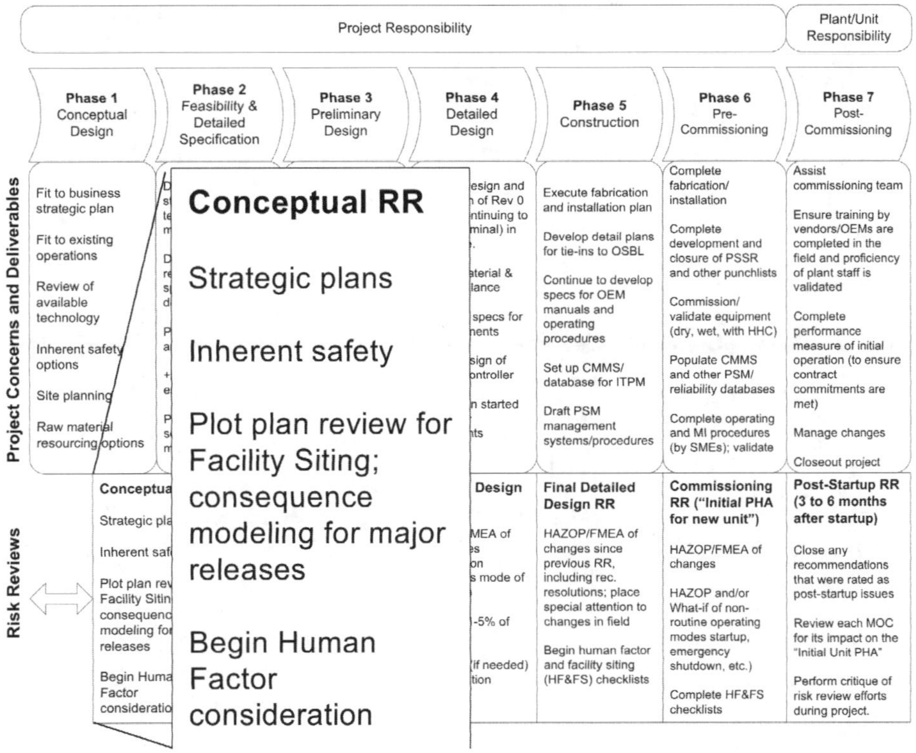

16

Keys to Excellent Risk Reviews in Conceptual & Preliminary Phases

- **On the risk review team, include:**
  - ➢ **senior operator from an existing or very similar unit**
  - ➢ **senior process engineer from an existing or similar unit**
  - ➢ **even if these persons must be contracted or must travel a great distance to attend the Risk Review**
- **Exclude project management from the PHA team meetings (they will have a chance later to accept/reject recommendations)**
- **Allow contingency in budget and schedule for possible changes**

17

Keys to Excellent Risk Reviews in Conceptual & Preliminary Phases (cont.)

- **NEVER let the contractor/vendor manage the risk review or provide the risk review leader. The leader should be:**
  - ➢ Independent of the project team
  - ➢ Independent of contractors/vendors
  - ➢ Independent of the Unit/Process/Plant that the major project is related to
  - ➢ Fully capable PHA team leaders/facilitators (well trained and practiced in the HAZOP, FMEA, and What-If methods)
- **Force the consideration of inherently safer & more reliable alternatives**

18

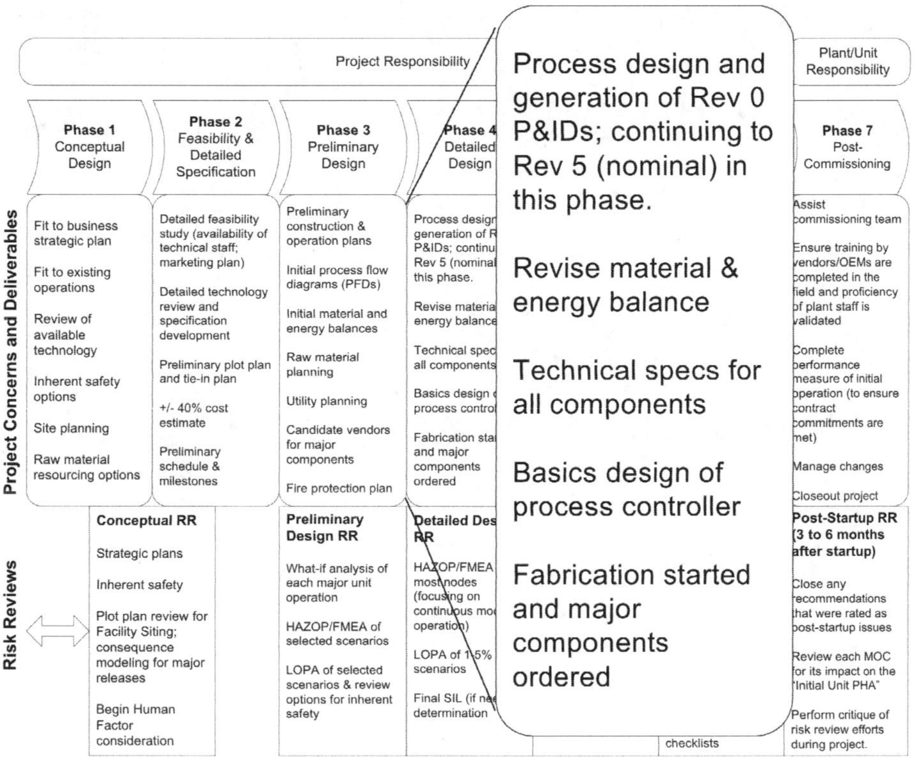

19

PSM Deliverables – Phase 3
Preliminary Design

- **Process safety information, including codes & standards, PFDs, thermal/kinetic chemistry information, material & energy balances, and materials of construction**
- **Facility siting basis set**
- **Begin emergency response planning**
- **Baseline info for future PHA**
- **Baseline info for future MI**
- **Continue employee participation**

20

PSM Deliverables – Phase 4
Detailed Design

- **Process safety information,** including P&IDs, revised materials of construction, safety interlocks and controls, equipment design basis and some final equipment details, 3D drawings

- Multiple layers of protection
- Detailed info for future PHA
- Detailed info for future MI
- Begin detailed emergency response plans
- Continue employee participation

21

PSM Deliverables – Phase 4
Detailed Design (continued)

- Maximize inherently safer design in the selected process, such as:
  - Lower feed pump pressures to make hydraulic overpressure impossible
  - Optimize reactor conditions and recycle loops to minimize waste
  - Design manual charge stations at ergonomic work height
  - Design tanks to withstand maximum possible pressure

22

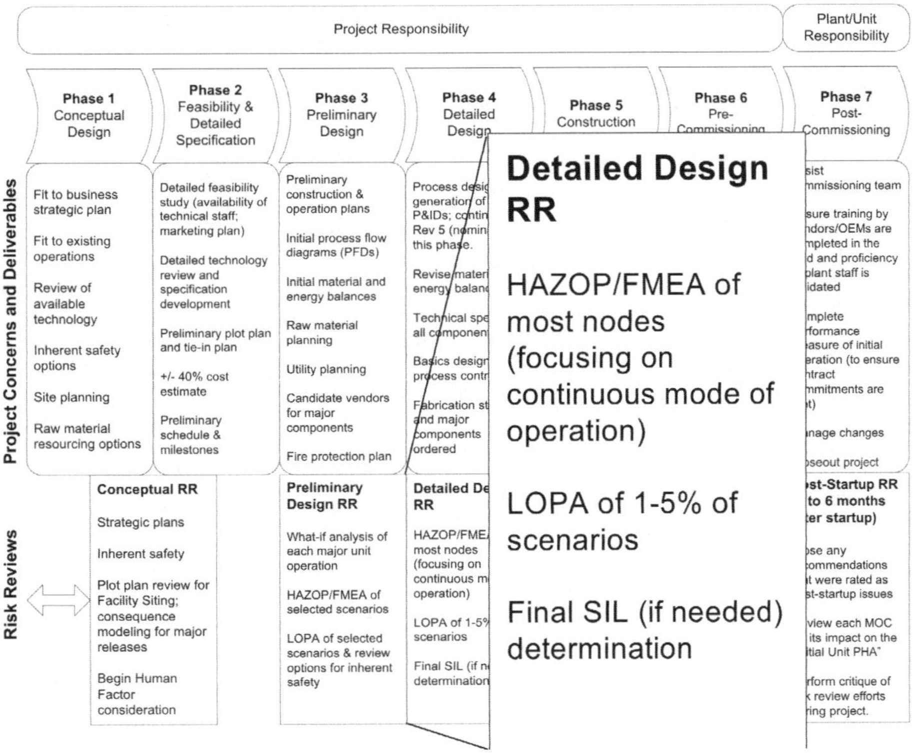

Keys to Performing Excellent Risk Reviews in Detailed Engineering Phase

- **Continue to have your most senior operators and process engineers on the Risk Review team**
- **Catch design problems before they create operational traps, by applying equal focus to hazards and operability/quality issues**

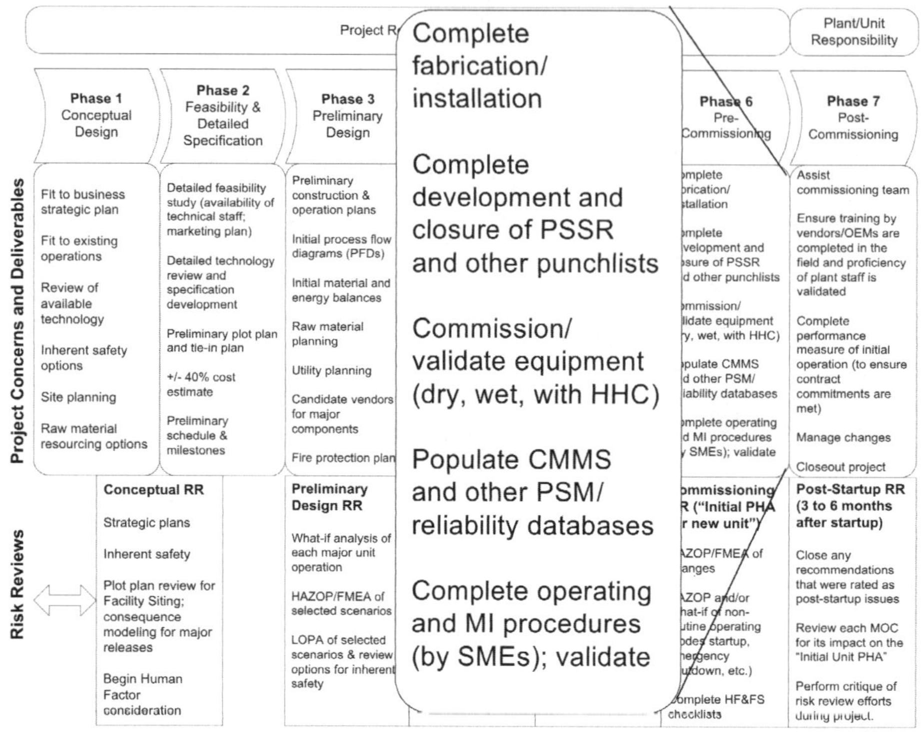

25

PSM Deliverables – Phase 5 Final Design

- Process safety information (revisions)
- Multiple layers of protection (revisions)
- Revisions/updates building to initial PHA
- Revised data for MI, including development of inspection, test, PM plans and populating databases
- Begin detailed consideration of human factors
- Data for developing operating and maintenance procedures
- Continue employee participation

26

PSM Deliverables – Phase 6 Commissioning

- Process safety information (revisions)
- Complete the initial PHA
- Revise MI plans, procedures, and database
- Complete detailed consideration of human factors
- Complete development of emergency response plans, operating procedures, and initial training
- Develop remaining elements of PSM, including MOC procedures and incident reporting and investigation system
- Conduct initial PSSR

27

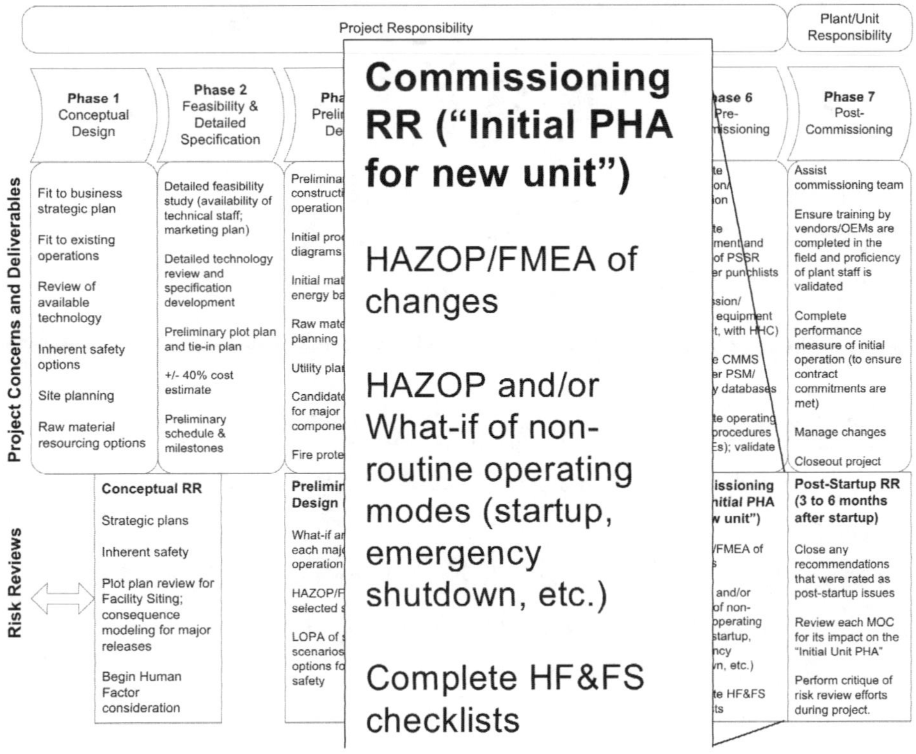

28

Key to Performing Excellent Risk Reviews in Commissioning Phase

- **Continue to have your most senior operators and process engineers on the Risk Review team, BUT add a second operator to the team**
- **Perform Risk Review (HAZOP/What-If) of start-up, shutdown, emergency shutdown, and on-line maintenance procedures ←** This is the most frequently observed weakness in the project risk review cycle *(see Chapter 9.1, Hazard Evaluation Procedures, 3rd edition, CCPS/AIChE 2008)*

29

Project is Complete, including:

- Mechanically complete and verified to be fit for duty
- Staff ready to operate and maintain equipment

- Initial PHA is complete (the risk review at each project phase has built sequentially to this deliverable)
- All PSM high level systems & implementation level documents, activities, and support systems are in place

30

Example: New Ethylene Plant

- Size: 250 nodes of equipment (nodes are vessels, columns, heaters, lines/exchanger circuits, etc.) plus all operating procedures (done in time for last project hazard review phase)

	HAZARD REVIEW ESTIMATES					
	Conceptual	Preliminary	Detailed	Construction	Pre-Startup	Total
Meeting Time	1 week	3 weeks	5 weeks	3 weeks	3 weeks	15 wks
Team Size	8	8	8	6	6	
Staff Hours Total	164	1152	2120	864	984	5284
Cumulative Schedule	2 months	6 months	14 months	18 months	22 months	24-30 mon
Focus	Plot plans, process options, tie-in options (WI and some modeling)	Cursory HAZOP, FMEA, and WI	Rigorous HAZOP, FMEA, and WI; start checklists	Changes since previous; finish checklists	SOP Hazard Review & Changes since previous	

32

Example: New Ore Conveying

- Size: 10 nodes of equipment (nodes are conveyors, screen, crushing, storage, dedusting, etc .) plus all operating procedures (done in time for last project hazard review phase)

	HAZARD REVIEW ESTIMATES		
	Detailed	Pre-Started	Total
Meeting Time	2 days	2 days	4 days
Team Size	10	10	
Staff Hours Total	218	202	419
Cumulative Schedule	3 months	9 months	10-11 mon
Focus	Rigorous WI, some FMEA; start checklists	SOP Hazard Review & Changes since previous	

33

Conclusions – Keys to Success

- Effective company leadership and effective project management, where the **primary goal** is an efficient operating unit (long-term process reliability and safety) – *where this primary goal outweighs the secondary goals of bringing in a project on schedule and on budget*
- Including experienced operators and experienced process engineers on the risk review teams, from the very start of the project
- Developing PSM elements at each project phase to better ensure the final process is efficient and safe to operate

34

Conclusions – Keys to Success (continued)

- Having subject matter experts develop procedures (operating, maintenance, and lab procedures) and performing the **risk review** of each of these procedures to ensure there are sufficient safeguards to protect the new process & people when the imperfect humans do not follow the procedures perfectly (1/100 error rate is typical best case)
- Developing trouble-shooting guides from PHA
- Delivering a project that allows sustainable control of human error

35

Thank you

Questions & Comments are Welcome

adakheel@piii.com

Stop by PII booth

36

An Approach to Improve Process Safety Management Audits

- Conference: CCPS Middle East Conference on Process Safety
- Year: 2017
- Time:
  Wednesday, October 11, 2017 - 10:00am-10:30am

The status of a process industry facilityâs risk-based process safety management program implementation and performance depends on many interconnected factors. Among the many factors that influence the status of implementation and performance are: maturity of adopted process safety management system, facility size and age, severity of process hazards, organizationâs safety culture, incidents history, risk tolerance, competency, and the regulatory requirements. The success of an organizationâs safety program in achieving the organizationâs own safety targets is typically measured utilizing a structured audit program. The used audit techniques may vary depending on the life cycle stage of the facility. Therefore, it is important to select the right audit technique that helps identify risk-based process safety gaps and establish basis for corrective actions.
Classical audit techniques have been focused on reviewing organizationâs written process safety management policies and procedures for completeness, interviewing stakeholders for awareness and understanding, and checking process safety records for degree of implementation.

However, to improve the effectiveness of an audit exercise, the process safety management program is further evaluated via assessing the strength of the process preventive and mitigative barriers, which is provides a view of the process safety management system performance in action.

In this presentation, an approach to conduct hazard identification and risk assessment in conjunction with process safety audit is described using the bow-tie technique. This approach emphasizes on hazard identification and determining the preventive and mitigative barriers. The strength of these barriers helps to evaluate the process safety performance at the site.

Ultimately this approach, this will allow identifying the high risk hazard and determine the performance of barriers and to showcase the ultimate consequences. Itâs pictorial representation of hazard, barriers, and consequences. PSM auditors use this tool to check the effectiveness of the barriers within the domain of organizations PSM element.

Risk-Based Management of Mechanical Integrity in Air Products Hydrogen Steam Methane Reformer Plants

Presentation by

Matthew Akhurst

Operations Director: Middle East, India, Egypt, and Turkey

October 11, 2017

AIR PRODUCTS

Agenda

Air Products Global H2 Capabilities
Risk Based Inspection Overview
Case Studies:

1. Hydrogen Purification
2. Steam Methane Reformer Tube Inspection
3. Steam Methane Reformer Outlet System

Questions & Answers
Appendix Case Study: CO2 Absorber Column IOW Shift

AIR PRODUCTS

Air Products Global Hydrogen Capabilities and Experience

- Worldwide leadership position in outsourced hydrogen production & recovery
  - Supply 3,000 MMSCFD of H2
  - Over 1600 operating years of experience
  - Maintained 40% market share for over 20 years
  - Strong focus on refining and chemical industries
  - Service Air Products and customer hydrogen assets
- Established reputation for high reliability operation
  - Strong pipeline positions
  - Operational excellence at plant level
- Own & operate over 100 hydrogen plants globally
  - 24+ world class high reliability SMR's
  - Operate ~10,000 tubes

AIR PRODUCTS

Risk Based Inspection

- A systematic process for factoring risk into decisions concerning how, when, and where to inspect pressure systems equipment, tanks and piping systems.
- Risk is defined using a criticality rating process, that combines the likelihood and the consequence of failure of equipment and piping systems
- Consequence of failure includes evaluation of safety, environmental, and financial impact
- Likelihood of failure is determined by identifying potential damage mechanisms based on mechanical design of the system and the system's Integrity Operating Window

Downtime Task? (Y/N)	PSM (Y/N)		ESCC	ISCC	CUI	ICML	EGCML	FC	General Visual (ref 01.20.37	OTHER	Notes	Safety/ Environmental	Economic	Failure Potential	Total Risk	Risk Ranking	M.I. or P.M.?	Task 1	Frequency 1	Task 2	Frequency 2
20	21	#	23	24	25	26	27	28	29	30	31	32	33	34	35	36	37	38	39	40	41
Y	N				X				X			3	2	2	12	H	MI	XXX	XX		

AIR PRODUCTS

Consequence of Failure

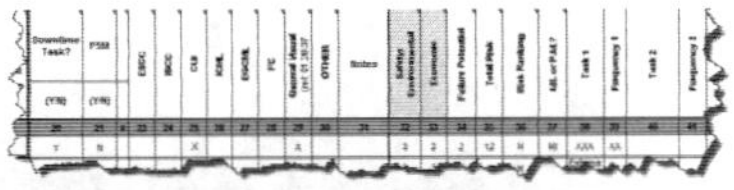

Type	Consideration
Safety	Process and Personnel safety, off site consequences
Environmental	Potential to approach or exceed permit req.
Financial	Cost to repair, lost sales, contract penalties, loss of goodwill

- All cases presented today share common consequence of failure
  - Release of flammable, toxic material
  - Potential PV Energy based on failure mode
  - Reduced capacity / likely unplanned outage
  - Significant repair costs
  - *Impact on Customer. Air Products works closely with our "over the fence" customers and know the criticality of hydrogen product to their operation and the consequences of unplanned reductions in supply.*

Likelihood of Failure

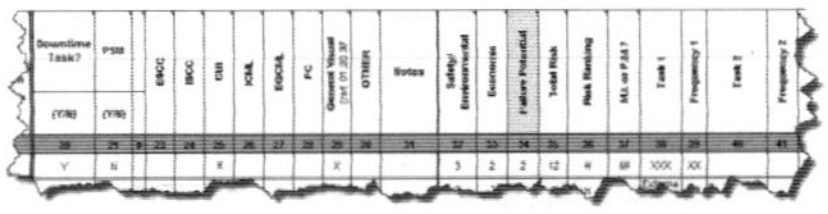

- Damage mechanisms are identified by evaluating mechanical design considerations and the process's Integrity Operating Windows (IOW)
- Mechanical Design considerations include material of construction, design basis, and fabrication details
  - Design basis considerations include
    - Creep, corrosion allowance, fatigue
  - Fabrication details include
    - Weld procedure, stress relief, weld contour, vessel shape
- Integrity Operating Windows (IOW) include both normal and transient conditions (startup and shutdown)
  - IOWs should be periodically validated to ensure changes have not introduced the potential for new damage mechanisms

Case Study - Hydrogen Purification

- Pressure Swing Adsorber (PSA) to purify Syngas using an adsorbent bed to remove impurities at process pressure, regenerated at low pressure (~5min cycle).
- Multiple vessels are required due to the adsorbent's limited capacity, requiring periodic regeneration

  AIR PRODUCTS

Hydrogen Purification

Probability of Failure

- Damage Mechanism Assessment
  - For applications operating below 200°C, fatigue is the Primary Damage Mechanism
- Likelihood of Fatigue Damage starts with Design and Fabrication
  - Design
    - Vessel designed for fatigue?
    - Design includes fabrication tolerances?
    - Fatigue design consistent with the actual IOW?
  - Fabrication
    - Roundness of vessel, particularly at longitudinal seams, can greatly impact local stresses
    - Weld quality can increase local stresses and create crack initiation points

  AIR PRODUCTS

Hydrogen Purification
Inspection Strategy

- Inspection strategy should focus on fatigue damage at weld locations
  - Both full penetration and attachment welds
- Likelihood and location of crack formation is strongly influenced by the vessel roundness at the weld seams (peaking) and by weld quality
  - Validating vessel shape during factory acceptance can increase the effectiveness of the inspection program as shape impacts whether a crack will likely initiate on the ID or OD vessel surface
- Timing of inspection is a function of the Integrity Operating Window
  - Condition Monitoring by tracking pressure cycle size (ΔP) and frequency allows inspection optimization

  AIR PRODUCTS

Hydrogen Purification
Fabrication – Weld Seam "Peaking"

- Vessel shape at longitudinal weld joint must be considered, and can add significant local stresses

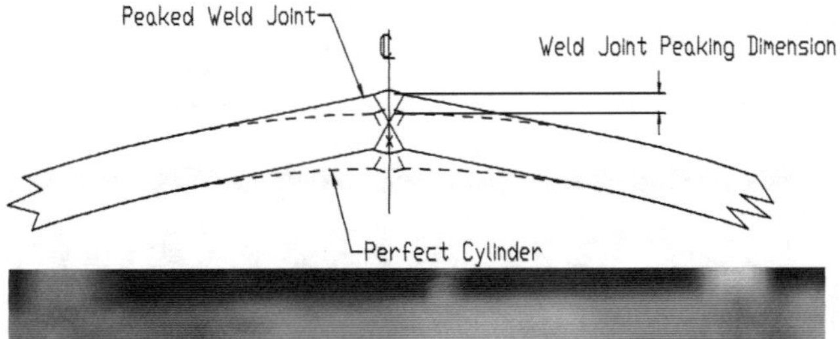

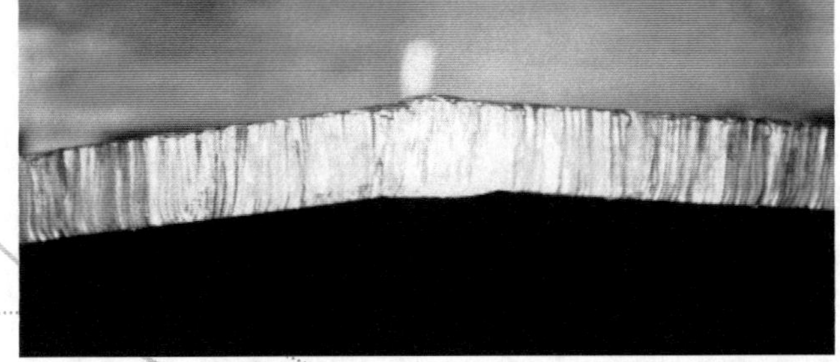

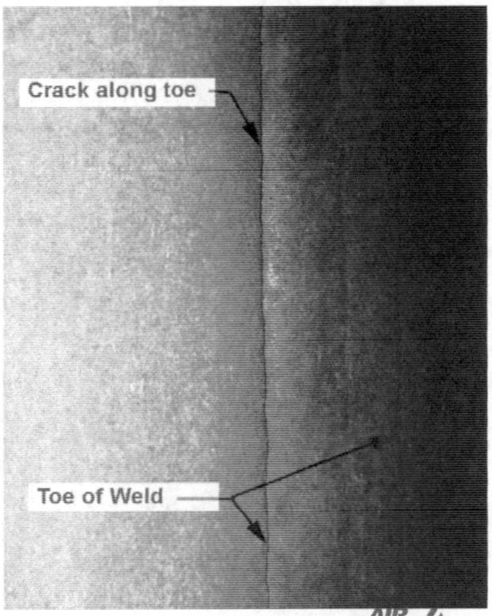

  AIR PRODUCTS

Hydrogen Purification
Design & Fabrication

- Attachment welds create local stress that when placed near longitudinal seams increase potential for fatigue damage

Hydrogen Purification
Inspection Strategy

- The first fatigue inspection typically is executed at the vessel's fatigue design half life
- Common methods of inspection include Ultrasonic Angle Beam Inspection (UTSW), Phased Array, and Magnetic Particle Surface Examination (MT)
- Future inspection requirements can be evaluated using a Fitness For Service (FFS) approach
  - API 579 Part 9 and BS7910 provide FFS guidance
- Industry Guidance Documents
  - CGA H-13 "Hydrogen Pressure Swing Adsorber (PSA) Mechanical Integrity Requirements"
  - API 571 "Damage Mechanisms Affecting Fixed Equipment in the Refining Industry"

Case Study - Steam Methane Reformer (SMR) Tube Inspection

- Typical large reformers contain hundreds of tubes
- Tubes designed to API 530 "Calculation of Heater-tube Thickness in Petroleum Refineries"
- Tubes are centrifugally cast and typically made out of HK40, HP50, or HP MicroAlloy

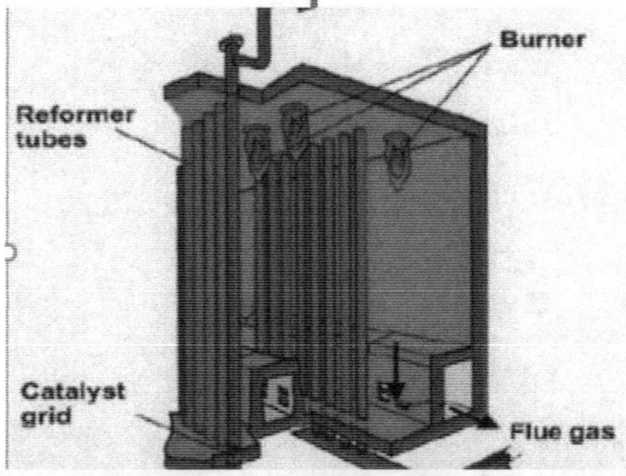

AIR PRODUCTS

SMR Tube Inspection
Damage Mechanism Assessment

- Primary Failure mechanisms are short term overheat damage, structural deformation, long term creep damage
- Short term overheat from flame impingement and catalyst issues
- Structural deformation (tube bowing) from improper support and uneven heating
- Long term creep from service
  - Tube temperature has large impact on creep life

AIR PRODUCTS

SMR Tube Inspection
Inspection Strategy

- Combination of on-line and off-line inspections
- On-line inspections
  - Burner Firing
  - Tube Temperature Measurement
- Off-line inspection
  - Visual inspection
  - Tube Crawling Creep inspection
  - Destructive Testing
- Industry Guidance
  - API 573 "Inspection of Fired Boilers and Heaters"

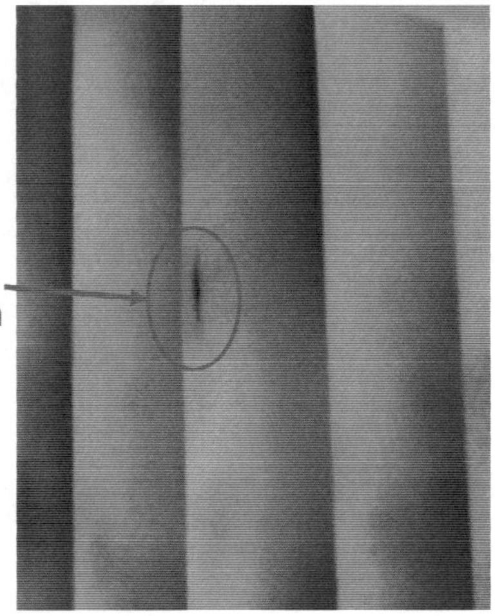

  AIR PRODUCTS

SMR Tube Inspection
Short Term Overheat

- Flame impingement

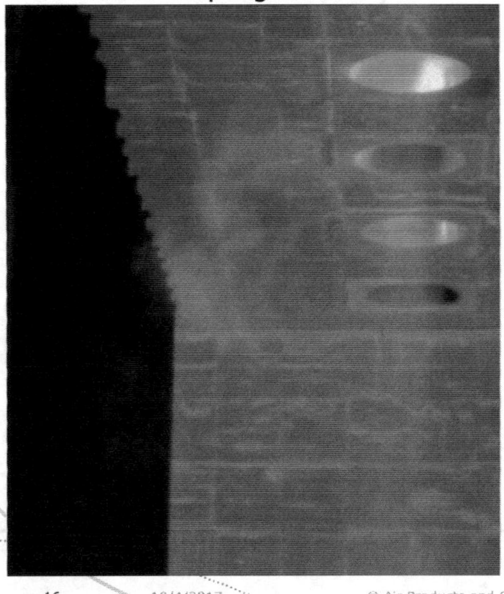

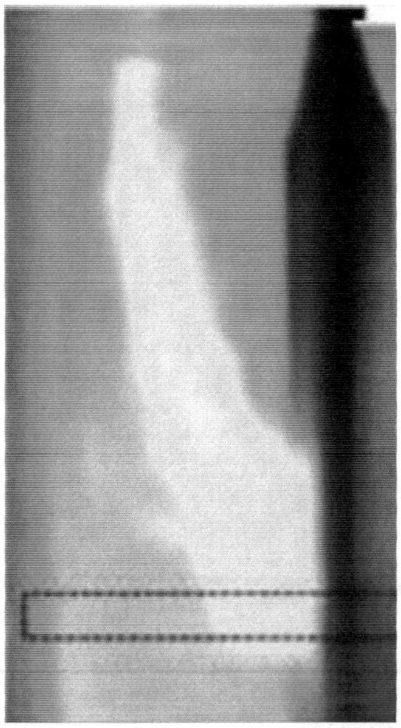

  AIR PRODUCTS

SMR Tube Inspection
Short Term Overheat

- Catalyst issues

17 10/4/2017 © Air Products and Chemicals, Inc. 2017 AIR PRODUCTS

SMR Tube Inspection
Long Term Creep

- Long Term Creep Damage occurs at the consistently hottest part of the tube, typically near the floor on a top fired, down flow furnace
- Creep fissures develop mid-wall and propagate to failure

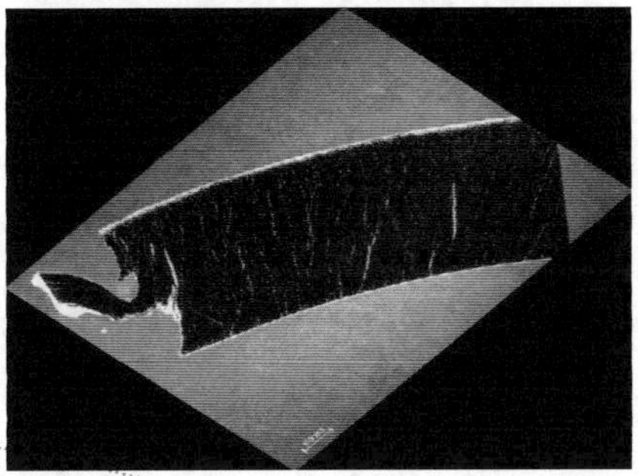
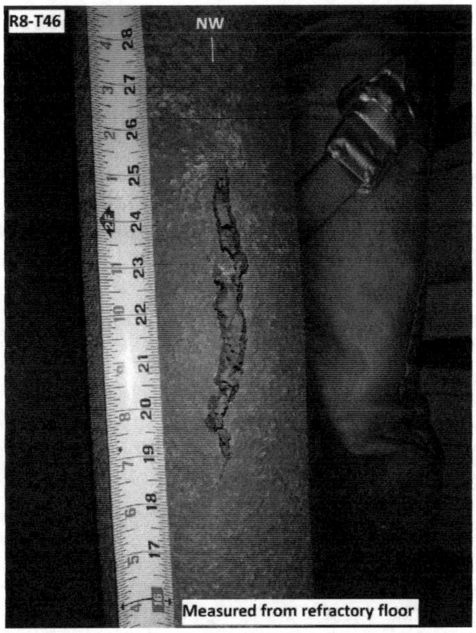

18 10/4/2017 © Air Products and Chemicals, Inc. 2017 AIR PRODUCTS

SMR Tube Inspection

- Visual inspection during operation
- Tightly balanced tube temperatures keep tubes within IOW
- Periodic tube temperature measurement via pyrometer
  - Correction factors must be applied to account for reflected energy error
    - Under-correction results in economic penalty
    - Over-correction can result in tube overheat & reduced tube life

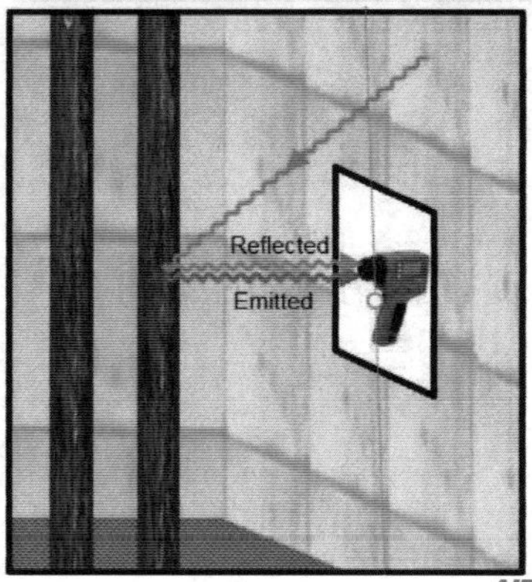

  AIR PRODUCTS

SMR Tube Inspection

- Tube temperature measurement/balancing using AP Proprietary digital imaging and modelling.
- Periodic tube temperature measurement via OmniCam™
  - OmniCam™ is an Air Products developed technology to measure tube temperature. Identifies hot tubes in locations not visible using standard technology.

  AIR PRODUCTS

SMR Tube Inspection

Tube Temperature Imaging & Balancing

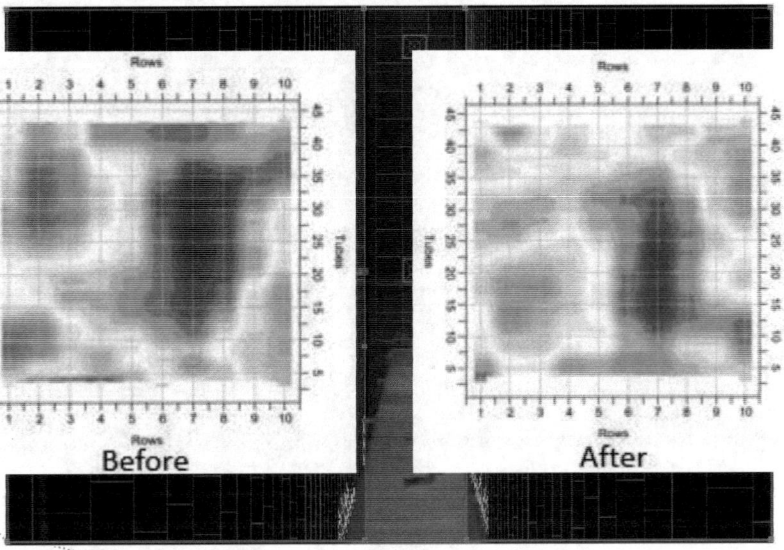

SMR Tube Inspection

Off-Line Inspection

- Reformer tube inspection via robotic crawler for creep damage
  - Crack formation
  - Diametrical growth
  - Tube Bowing
- Remaining tube life via destructive testing

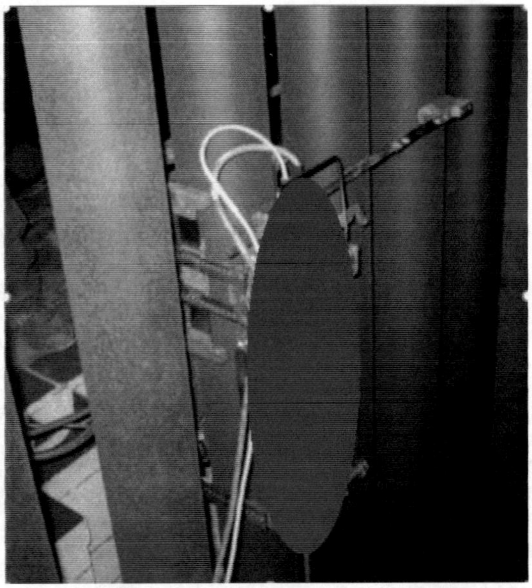

Case Study - SMR Hot Outlet System Inspection

- Comprised cast High Nickel Alloy headers, Inco 800HT wrought pigtails, and low alloy refractory lined piping that conveys ~30 barg, 870 C Syngas to downstream unit operations
- Typically designed with Short or Long Pigtail Design

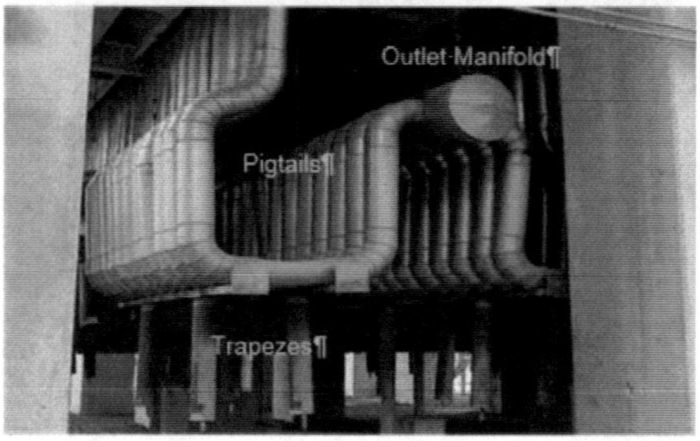

  AIR PRODUCTS

SMR Hot Outlet System Inspection

- Primary damage mechanisms are:
  - Short term stress overload from improper support or refractory failure
  - Long term creep damage.
  - HTHA possible if internal refractory lining degrades, allowing metal temperature to enter HTHA risk zone.

  AIR PRODUCTS

SMR Hot Outlet System Inspection

- On-line inspection is comprised of visual inspection and infra-red thermography
- Off-line inspection is comprised of creep inspections and visual inspection of refractory lined components
- Industry Guidance Documents
  - CGA H-12 "Mechanical Integrity of Syngas Outlet Systems"
  - API 571 "Damage Mechanisms Affecting Fixed Equipment in the Refining Industry"
  - API 573 "Inspection of Fired Boilers and Heaters"
  - API 941 "Steels for Hydrogen Service at Elevated Temperatures and Pressures in Petroleum Refineries and Petrochemical Plants"

AIR PRODUCTS

SMR Hot Outlet System Inspection
On-Line Inspection

- Visual inspection of spring hangers, supports, and outlet system

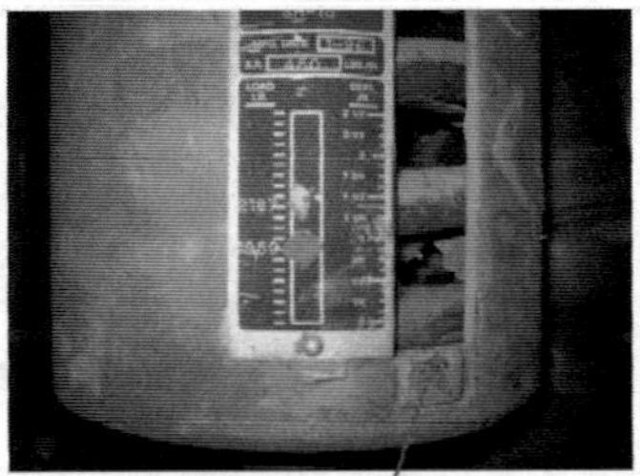

Spring Hanger bottomed out (no load)

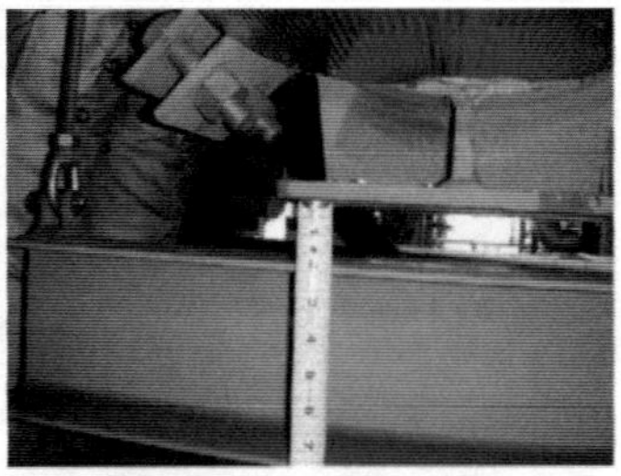

Gap at Support

- Visual inspection for damaged insulation that can cause high thermal stress

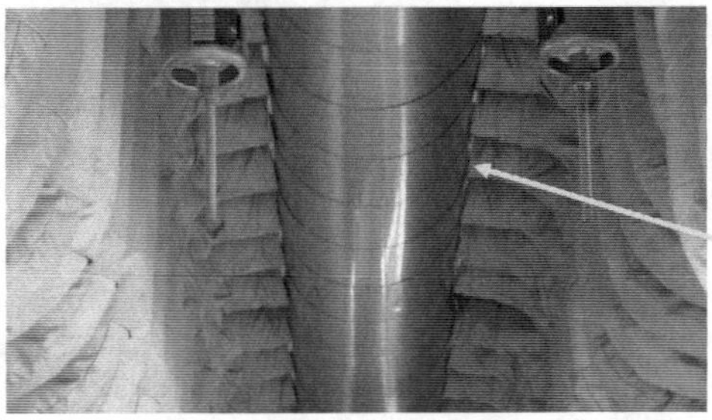

Insulation gap creates thermal stress

AIR PRODUCTS

SMR Hot Outlet System Inspection
On-Line Inspection

- IR Thermography of Refractory Line Equipment

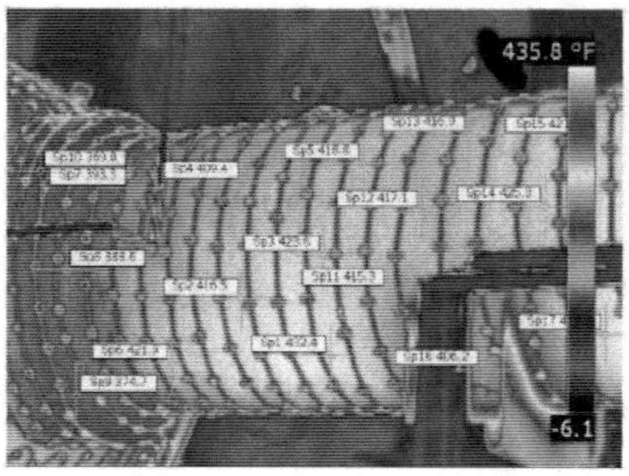

  AIR PRODUCTS

SMR Hot Outlet System Inspection
Off-Line Inspection

- Invasive visual inspections of supports

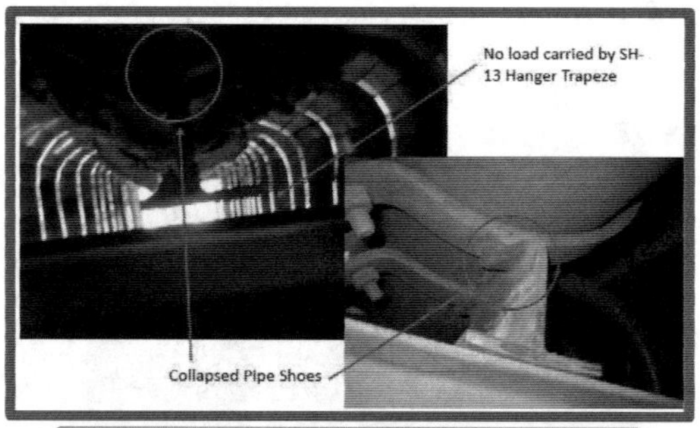

- Creep inspection via diametrical growth of pigtails

  AIR PRODUCTS

SMR Hot Outlet System Inspection
Off-Line Inspection

- Invasive visual inspections of supports

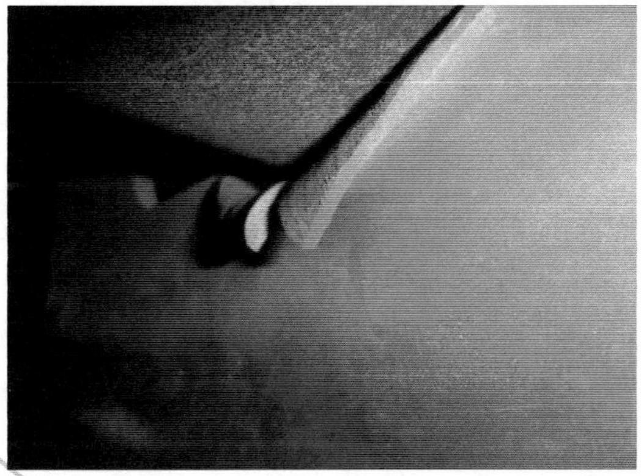

SMR Hot Outlet System Inspection
Off-Line Inspection

- Surface and Radiographic Inspection of Cast Outlet System Welds

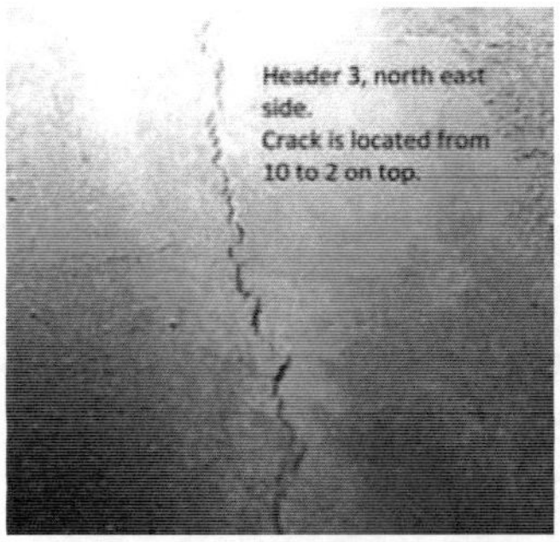

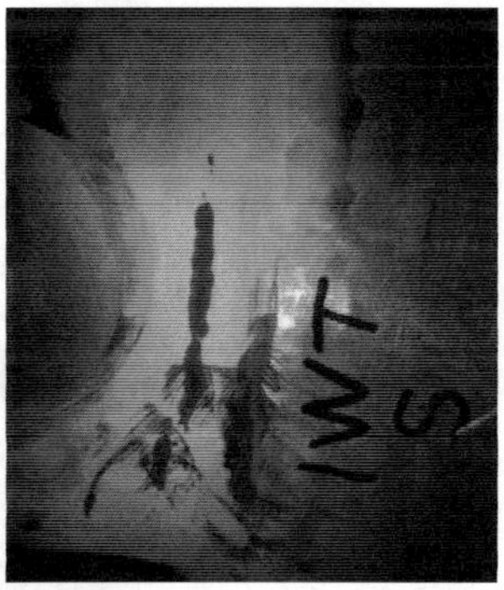

- Internal Visual Inspection of Refractory Lined Equipment

Questions?

31 10/4/2017

AIR PRODUCTS

Thank you
tell me more

AIR PRODUCTS

32 10/4/2017

Case Study - Process Column – IOW Shift

- CO2 removal service using amine technology
- 20 m tall, CS PWHT Column operates at ~ 40 barg and ambient temperature in Syngas service
- Primary damage mechanism is internal corrosion, which is managed through the use of corrosion inhibitors

33 10/4/2017 © Air Products and Chemicals, Inc. 2017 AIR PRODUCTS

Process Column – IOW Shift

- Inspection strategy used a combination of Condition Monitoring, Internal Inspection, and Corrosion Monitoring
- Condition monitoring periodically sampled process for inhibitor and iron levels.
- Sampling showed drop in inhibitor level and increase in iron levels

34 10/4/2017 © Air Products and Chemicals, Inc. 2017 AIR PRODUCTS

Process Column – IOW Shift

- Response to IOW shift was to restore inhibitor levels and alert Inspection group
  - Corrosion monitoring accelerated
- Condition monitoring showed restoration of inhibitor levels, but continued elevated iron levels.
  - Column passivation required, but outage window not available
  - Corrosion Monitoring strategy modified to ensure safe operation

  AIR PRODUCTS

Process Column – IOW Shift

- Corrosion monitoring adjusted to focus on bottom column in areas impacted by inlet distributor
  - Impingement created accelerated corrosion rates
  - API 579 FFS was employed to continue operation of the column until a repair window was available

Elev	1'	2'	3'	4'	5'	6'	7'	8'	9'	10'	11'	12'	13'	14'	15'	16'	17'	18'
Design	[illegible]	[illegible]	[illegible]	[illegible]	[illegible]	[illegible]	[illegible]	[illegible]	[illegible]	[illegible]	[illegible]	[illegible]	[illegible]	[illegible]	[illegible]	[illegible]	[illegible]	[illegible]
Tmin	[illegible]	[illegible]	[illegible]	[illegible]	[illegible]	[illegible]	[illegible]	[illegible]	[illegible]	[illegible]	[illegible]	[illegible]	[illegible]	[illegible]	[illegible]	[illegible]	[illegible]	[illegible]
A (13')	[illegible]	[illegible]	[illegible]	[illegible]	[illegible]	[illegible]	[illegible]	[illegible]	[illegible]	[illegible]	[illegible]	[illegible]	[illegible]	[illegible]	[illegible]	[illegible]	[illegible]	[illegible]
B (12')	[illegible]	[illegible]	[illegible]	[illegible]	[illegible]	[illegible]	[illegible]	[illegible]	[illegible]	[illegible]	[illegible]	[illegible]	[illegible]	[illegible]	[illegible]	[illegible]	[illegible]	[illegible]
C (11')	[illegible]	[illegible]	[illegible]	[illegible]	[illegible]	[illegible]	[illegible]	[illegible]	[illegible]	[illegible]	[illegible]	[illegible]	[illegible]	[illegible]	[illegible]	[illegible]	[illegible]	[illegible]
D (10')	[illegible]	[illegible]	[illegible]	[illegible]	[illegible]	[illegible]	[illegible]	[illegible]	[illegible]	[illegible]	[illegible]	[illegible]	[illegible]	[illegible]	[illegible]	[illegible]	[illegible]	[illegible]
E (9')	[illegible]	[illegible]	[illegible]	[illegible]	[illegible]	[illegible]	[illegible]	[illegible]	[illegible]	[illegible]	[illegible]	[illegible]	[illegible]	[illegible]	[illegible]	[illegible]	[illegible]	[illegible]
F (8')	[illegible]	[illegible]	[illegible]	[illegible]	[illegible]	[illegible]	[illegible]	[illegible]	[illegible]	[illegible]	[illegible]	[illegible]	[illegible]	[illegible]	[illegible]	[illegible]	[illegible]	[illegible]
G (7')	[illegible]	[illegible]	[illegible]	[illegible]	[illegible]	[illegible]	[illegible]	[illegible]	[illegible]	[illegible]	[illegible]	[illegible]	[illegible]	[illegible]	[illegible]	[illegible]	[illegible]	[illegible]
H (6')	[illegible]	[illegible]	[illegible]	[illegible]	[illegible]	[illegible]	[illegible]	[illegible]	[illegible]	[illegible]	[illegible]	[illegible]	[illegible]	[illegible]	[illegible]	[illegible]	[illegible]	[illegible]
I (5')	[illegible]	[illegible]	[illegible]	[illegible]	[illegible]	[illegible]	[illegible]	[illegible]	[illegible]	[illegible]	[illegible]	[illegible]	[illegible]	[illegible]	[illegible]	[illegible]	[illegible]	[illegible]
J (4')	[illegible]	[illegible]	[illegible]	[illegible]	[illegible]	[illegible]	[illegible]	[illegible]	[illegible]	[illegible]	[illegible]	[illegible]	[illegible]	[illegible]	[illegible]	[illegible]	[illegible]	[illegible]
K (3')	[illegible]	[illegible]	[illegible]	[illegible]	[illegible]	[illegible]	[illegible]	[illegible]	[illegible]	[illegible]	[illegible]	[illegible]	[illegible]	[illegible]	[illegible]	[illegible]	[illegible]	[illegible]
L (2')	[illegible]	[illegible]	[illegible]	[illegible]	[illegible]	[illegible]	[illegible]	[illegible]	[illegible]	[illegible]	[illegible]	[illegible]	[illegible]	[illegible]	[illegible]	[illegible]	[illegible]	[illegible]
M (1')	[illegible]	[illegible]	[illegible]	[illegible]	[illegible]	[illegible]	[illegible]	[illegible]	[illegible]	[illegible]	[illegible]	[illegible]	[illegible]	[illegible]	[illegible]	[illegible]	[illegible]	[illegible]

  AIR PRODUCTS

ABS Group

Mechanical Integrity (MI) Role in Managing Aged Safety Critical Equipment (SCE)

An approach to Excellence in Process Safety

Presented by: Gary Graham
Vice President
Europe, Middle East & Africa

Osama Elsaadawy
Author

Agenda

1. Introduction & Background
2. What Is Ageing Equipment?
3. Identifying Ageing
4. Ageing Related Threats and Damage Mechanisms
5. Ageing Stages
6. Integrity Barriers & Safety Critical Equipment/Elements "SCE"
7. Example – How Barriers' Works
8. Examples indicate Ageing' Symptoms of SCEs
9. Overview of Management of Ageing SCE Considerations
10. Some Indicators (KPIs) of Ageing SCE
11. Assurance of SCE PS & Assurance Activities
12. How Ageing Affect SCEs Performance Standard
13. References

ABS Group

Introduction & Background

The existing infrastructure and assets within the oil and gas industry are continuing to age, resulting in increased risk of equipment failure and integrity uncertainty.

Many existing operating assets in the region - in particular in offshore locations - are 30 to 40 years old.

The management of plant ageing has in recent years become a topic of increasing significance, in relation to decreasing plant integrity threats, increasing certainty regarding safe operation, equipment reliability and integrity and overall effective PSM program.

Implementing a robust management system for Aged SCE is a sub-set of an overall effective plant Mechanical Integrity (MI) program (driven by the OSHA 1910.119 standard) to approach Excellence in Process Safety.

ABS Group

3

What Is Ageing Equipment?

- Ageing is not about how old your equipment is; it is about its condition, and how it is changing over time.
- Overall, Ageing equipment/plant is plant which is, no longer considered fully fit for purpose due to deterioration or obsolescence in its integrity or functional performance.
- The significance of Ageing relates to the potential effect on the equipment's Performance Standards (i.e. Functionality, Availability, Reliability and overall Safety)

ABS Group

Identifying Ageing

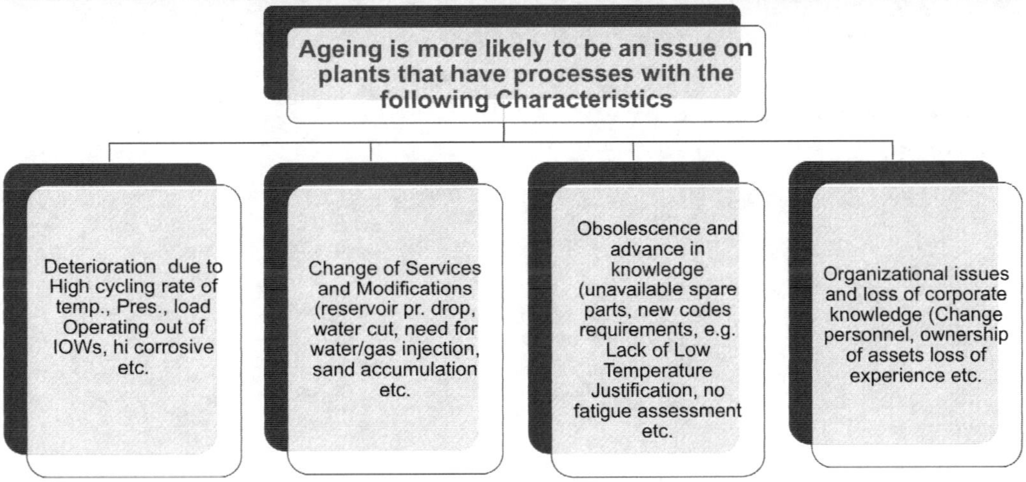

ABS Group

5

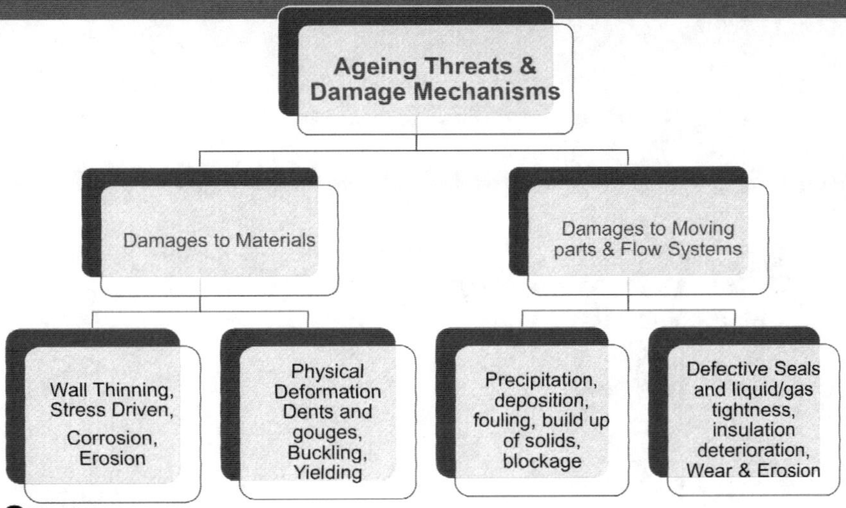

ABS Group

6

SCE Ageing Stages - Variation of accumulated damage during equipment service

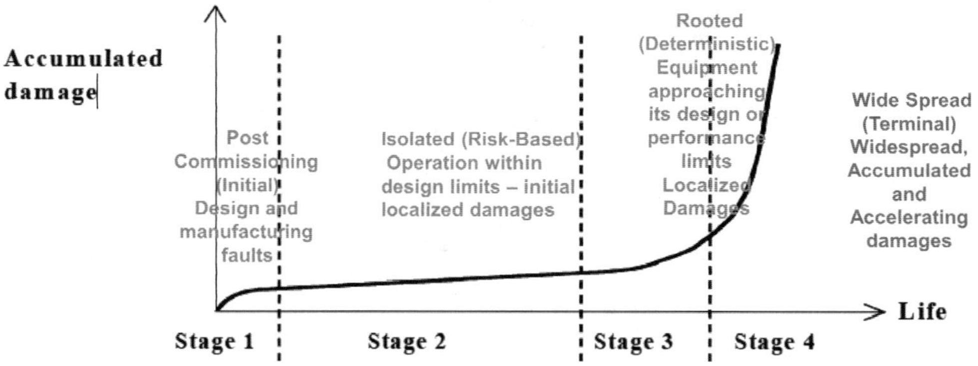

ABS Group

7

Integrity Barrier & Safety Critical Equipment/Elements "SCE"

Asset types susceptible to ageing can be segregated into the following Categories/Barriers:

- **Hardware Barriers** (all types of equipment can be susceptible to ageing mechanisms")
- **Operational Barriers** (MOC, Operating Procedures, Manuals, Guidelines, COPs etc.)
- **Organizational Barriers** (e.g. Organizational Structure, Policies, Mission & Vision, Corporate knowledge etc.)

8

Integrity Barrier & Safety Critical Equipment/Elements "SCE"

Barrier: Is defined as any technical (e.g. Hardware Barrier), operational or organizational measures whose function prevents, decreases the probability of hazardous events, or which limits the consequences of such events.

SCE: Is equipment or system (or part of it), failure of which could cause or contribute substantially to MAH, or the purpose of it is to prevent or limit the effect of MA.

Equipment integrity is ensured by the existence and performance of a series of integrity barriers (eight hardware barriers)

9

Integrity Hardware Barrier (Swiss Cheese) Diagram

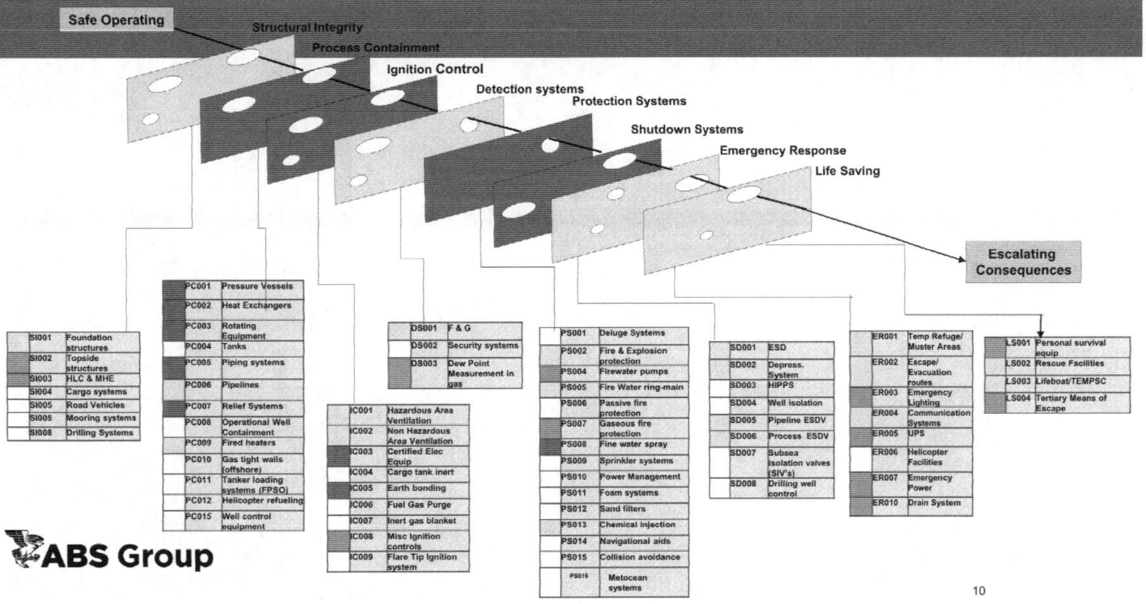

ABS Group

10

Integrity Barrier (Swiss Cheese) Diagram

- Small holes in hardware barriers, indicate a weaknesses or potential degradation (Ageing) of the barrier's performance.
- If those holes, between adjacent hardware barriers, lined up, there may not be effective barriers (prevention/protection etc.) in place between safe operations and escalating consequences, leading to a Major Accident.

ABS Group

How Barriers' Works

- The following slide illustrates an example of Ageing' effect on Barriers and their group of SCEs in their role in a Hazard Scenario
- **Example: Hydrocarbon Leaks, Fire & Explosion:**

ABS Group

Example: Hydrocarbon Leaks, Fire & Explosion:

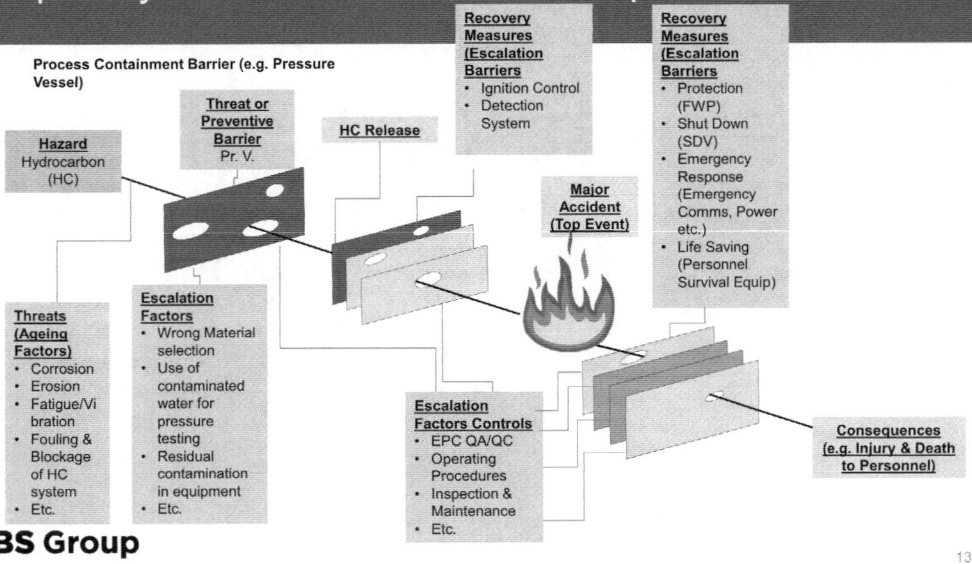

ABS Group

13

Ageing' Symptoms on SCEs

- These Barriers/SCEs are borne to ageing processes, that reducing efficiency in their role of controlling MAH.
- All of the above identified Barriers/SCE need regular inspection and maintenance to detect ageing symptoms and indicators in time.

ABS Group

14

The following are Examples indicate Ageing' Symptoms of SCEs:

	Examples indicate Ageing' Symptoms of SCEs
Hydrocarbon Containment	Internal & external corrosion
	Accelerated corrosion due to enhanced corrosivity of well fluid e.g. if H2S increased as well become depleted
	Blockage of well production tubing and process plant with scale
	Sand and water erosion/corrosion in bottom of J-tubes/risers
	Abrasion of coating on piping and equipment in the splash zone, leading to enhanced corrosion

ABS Group

15

The following are Examples indicate Ageing' Symptoms of SCEs:

	Examples indicate Ageing' Symptoms of SCEs
Detection & Control Systems (e.g. F&G detectors)	Corrosion of detectors, cable trays, Ex enclosures, electrical bonding points etc.
	Drift in detector response
	Poisoning of catalytic sensors
	Blockage of sintered metal screen in catalytic sensors
	Open path detectors blocked by obstructions such as scaffolding
	Obsolescence leading to non-availability of like-to-like spare parts and lack of vendor support
	Old application programs may be incompatible with current hardware

ABS Group

16

The following are Examples indicate Ageing' Symptoms of SCEs:

	Examples indicate Ageing' Symptoms of SCEs
Active Fire Protection (Deluge System)	Deluge system degradation and nozzle blockage due to corrosion
	Older systems may not be compatible with current standards
	Vulnerability to explosion damage
	Leaks in air trigger systems
	Wear and tear of pumps and valves
	Obsolete components
	Deterioration/dilution of foam concentrate

ABS Group

17

Overview of Management of Ageing SCE Considerations

- Management of Ageing SCE can be defined as the sum of all those tasks and processes which together contribute to ensuring that the progress of Ageing on a specific equipment or plant are monitored, with remedial action being taken as necessary, such that loss of containment of process fluids is avoided.
- Success in Managing Ageing SCEs is the start point for deciding the SCE' Life Extension.

ABS Group

18

Overview of Management of Ageing SCE Considerations

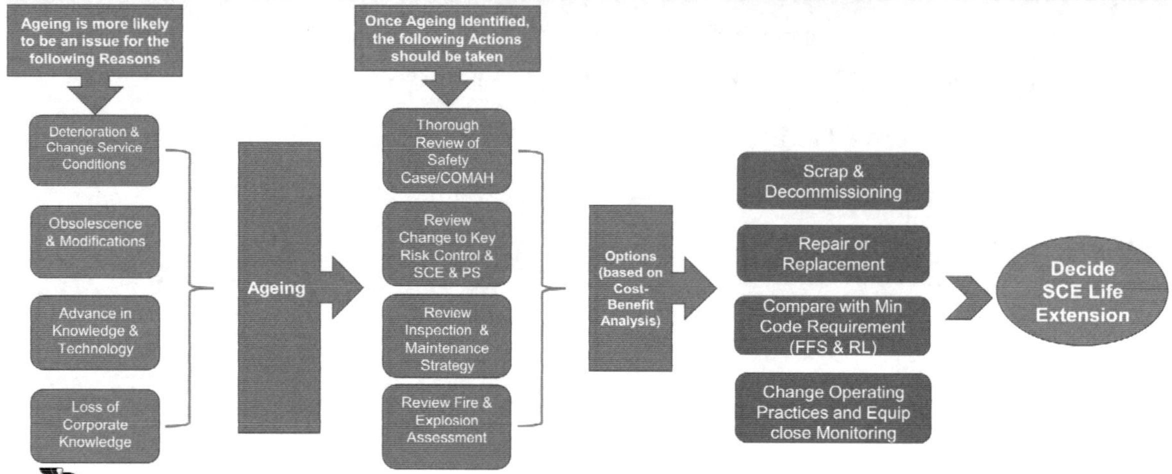

ABS Group

19

Some Indicators (KPI) of Ageing SCE

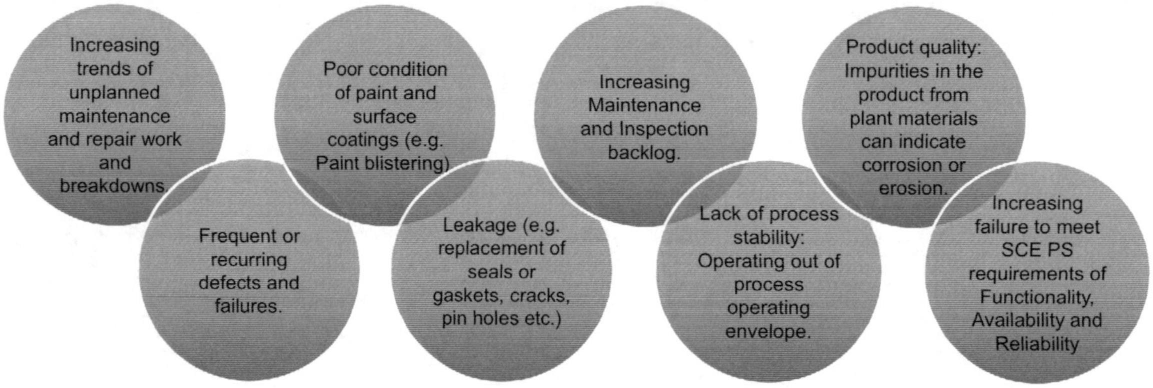

ABS Group

20

Assurance of SCE PS

- Assurance is all activities necessary to make sure that the SCE are suitable, have the desired integrity and will continue to perform their function safely.

Assurance Activities include:

- Activities such as design, construction, installation, commissioning, operation, modification, repair, inspection, testing and maintenance, that carried out by or on behalf of the asset owner / operator (known also as HSE Critical Integrity Activities).

ABS Group

21

How Ageing Affect SCEs Performance Standard

SCE - UPS			
FUNCTIONALITY			
Function	**Criteria**	**Assurance Activities**	**Updated Assurance Activities**
UPS shall maintain power to the defined emergency systems	• UPS systems shall provide a back-up power supply to enable continued operation of the following emergency systems in the event of failure of normal and essential power. • The following systems shall have a minimum duration of XX minutes. ➢ Shutdown Systems (XX Amperehours) ➢ Fire and Gas (XX Amperehours) ➢ Emergency comms & PA/GA Systems (XX Amperehours)	• Review Vendor Data Sheets for UPS Systems. • Review single line diagrams from ups to control system/ critical equipment • Review Inspection and test plans (ITPs) • Review FAT Records • Conduct periodic load live test	• Review/Revise the Inspection and Maintenance plans for UPS • Reschedule the periodic load live test (more frequent) based on new loads and new adjusted minimum time duration based on recent safety & risk assessment

ABS Group

22

How Ageing Affect SCEs Performance Standard

SCE – COMMUNICATION SYSTEMS			
AVAILABILITY			
Function	**Criteria**	**Assurance Activities**	**Updated Assurance Activities**
The PA/GA System shall maintain 100% Availability.	• The PA system has been designed with built in redundancy. • The PA/GA shall be a dual (sub- systems A & B) system, with each sub-system being capable of independently providing audible alarms, visual alarms and speech in all areas, such that no single failure can impair both sub-systems. Each sub- system shall be fed from independent Uninterruptible Power Supplies.	• Review Vendor Data Sheets & Specs • Review Telecom Equipment Layout Drawings • Review Inspection & Test Plan and Reports • Review Hazardous Area Classification Drawings • Conduct scheduled inspection and maintenance.	• Revise Hazardous Area Classification Drawings • Revise Telecom Equipment Layout Drawings based on new actual locations (inaudible location due to increasing noise) • Reschedule inspection and maintenance plans.

ABS Group

23

How Ageing Affect SCEs Performance Standard

SCE – ESD SYSTEM			
RELIABILITY			
Function	**Criteria**	**Assurance Activities**	**Updated Assurance Activities**
The entire Fire & Gas system shall meet defined Reliability.	• The Fire & Gas Detection system design and engineering shall be carried out and performed to meet requirements of SIL ratings as defined within IEC 61508. • F&G System processor/controller will be SIL? • F&G Loops in defined high risk areas will be SIL? • F&G Loops in defined low risk areas will be SIL? (ref to QRA study for target SIL	• Review Vendor Data Sheets for F&G Detectors & System • Review Cause and Effect Diagram for Review F&G Detectors Layout Drawings • Review Hazardous Area Classification Drawings • Conduct scheduled inspection and maintenance	• Revise Hazardous Area Classification Drawings • Revise F&G detectors Layout Drawings based on recent fire & explosion QRA • Reschedule inspection and maintenance plans

ABS Group

24

How Ageing Affect SCEs Performance Standard

SCE – JACKET STRUCTURE				
SURVIVABILITY				
Event	**Component**	**Criteria**	**Assurance Activities**	**Updated Assurance Activities**
Fire and Explosion	Primary structure within X metres of the YY deck. Pressure vessles ZZ, pipelines ESDV etc. Reference to Fire Risk Assessment.	• Passive Fire Protection is applied to structural members in the defined area • The fireproofing shall satisfy minimum 2-hours fire rating	• Review PFP zone layout and ensure requirements are as per the PFP schedule. • Review vendor document (PFP) provided and applied in accordance with specification • Conduct visual inspection and scheduled inspection and maintenance	• Revise PFP zone layout and ensure requirements are as per the PFP schedule based on recent fire assessment • Reschedule visual inspection and scheduled inspection and maintenance and include new identified PFP locations

ABS Group

25

ABS Group

THANK YOU! Time for Q&A

شركة الخليج لصناعة البتروكيماويات

Gulf Petrochemical Industries Company

Toward Reliability Excellence

2017 CCPS Middle East Process Safety Conference

GPIC in Brief

- Established in December 1979
- Joint-venture between the Government of Bahrain (The Oil & Gas holding company), Kingdom of Saudi Arabia (SABIC) & Kuwait (PIC)
- Utilises Natural Gas as a raw material
- Produces: Ammonia, Methanol and Urea
- Started production in 1985
- Production capacity: 1,500,000 tonnes annually
- Manpower: 481 employees

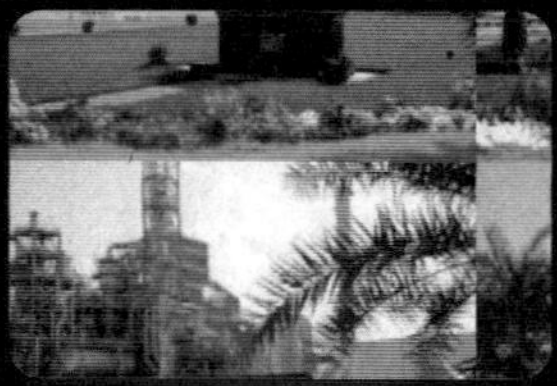

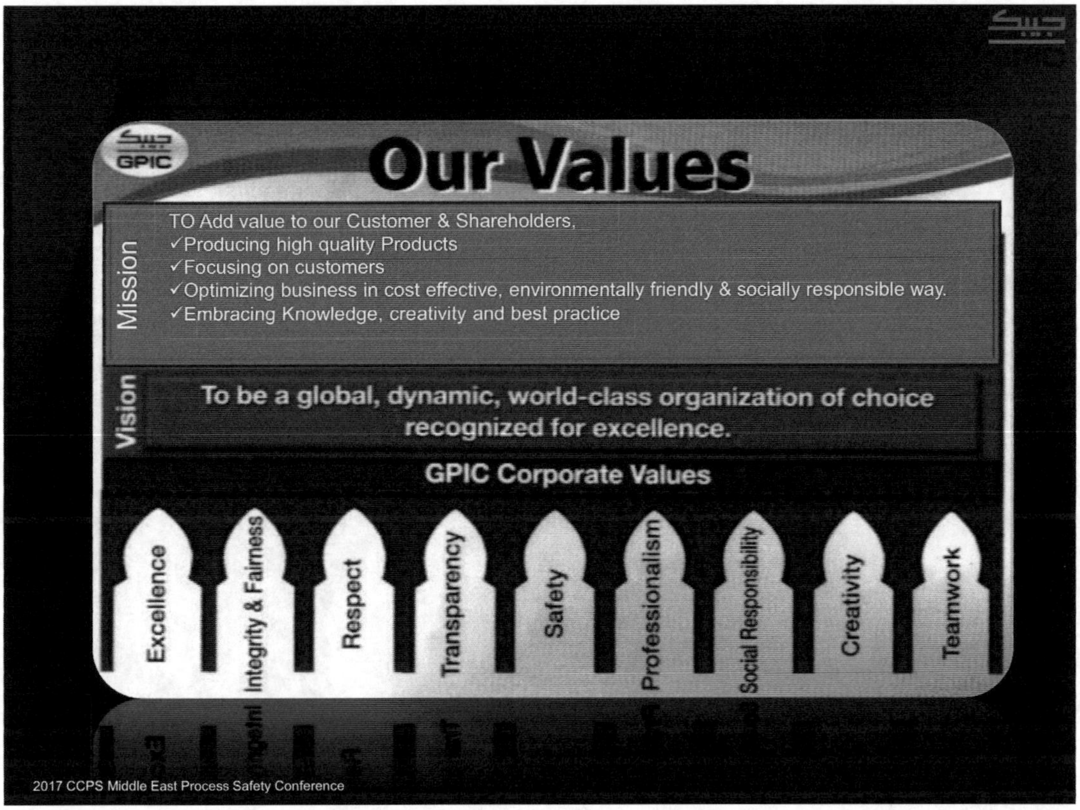
GPIC
Our Values
Mission
TO Add value to our Customer & Shareholders,
✓Producing high quality Products
✓Focusing on customers
✓Optimizing business in cost effective, environmentally friendly & socially responsible way.
✓Embracing Knowledge, creativity and best practice
Vision
To be a global, dynamic, world-class organization of choice recognized for excellence.
GPIC Corporate Values
Excellence
Integrity & Fairness
Respect
Transparency
Safety
Professionalism
Social Responsibility
Creativity
Teamwork
2017 CCPS Middle East Process Safety Conference

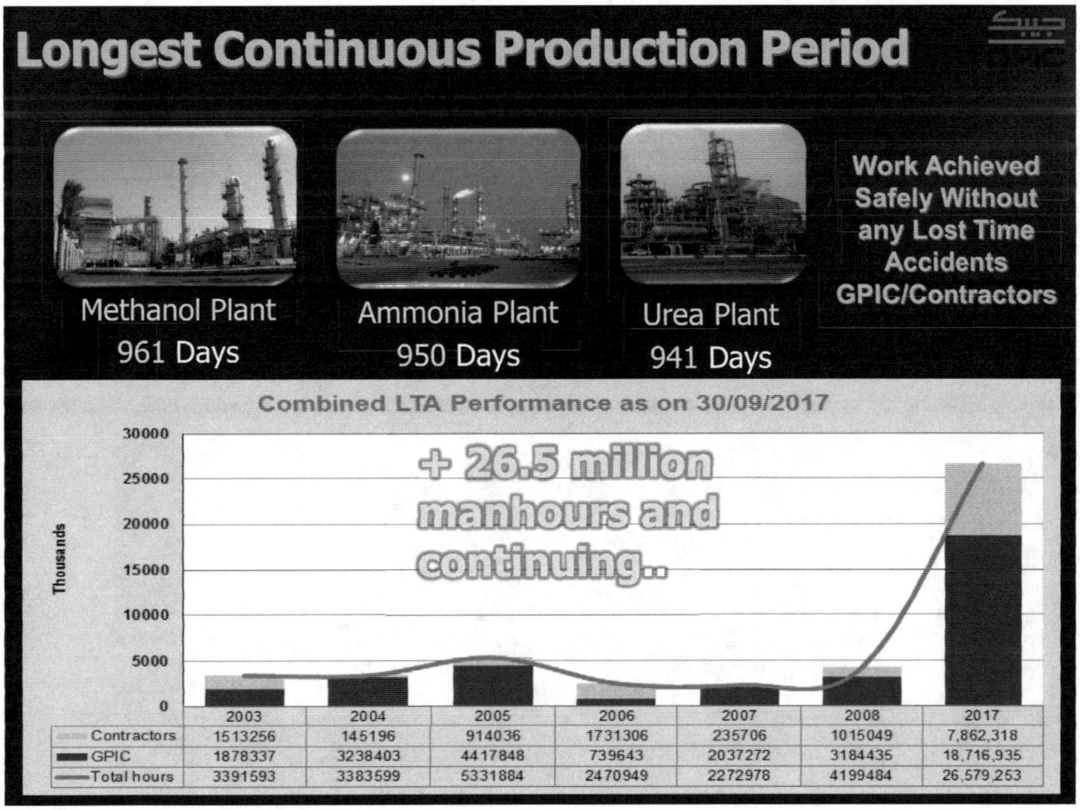
Longest Continuous Production Period
Methanol Plant
961 Days
Ammonia Plant
950 Days
Urea Plant
941 Days
Work Achieved Safely Without any Lost Time Accidents GPIC/Contractors
Combined LTA Performance as on 30/09/2017
+ 26.5 million manhours and continuing..
Thousands
30000
25000
20000
15000
10000
5000
0
2003 2004 2005 2006 2007 2008 2017
Contractors 1513256 145196 914036 1731306 235706 1015049 7,862,318
GPIC 1878337 3238403 4417848 739643 2037272 3184435 18,716,935
Total hours 3391593 3383599 5331884 2470949 2272978 4199484 26,579,253

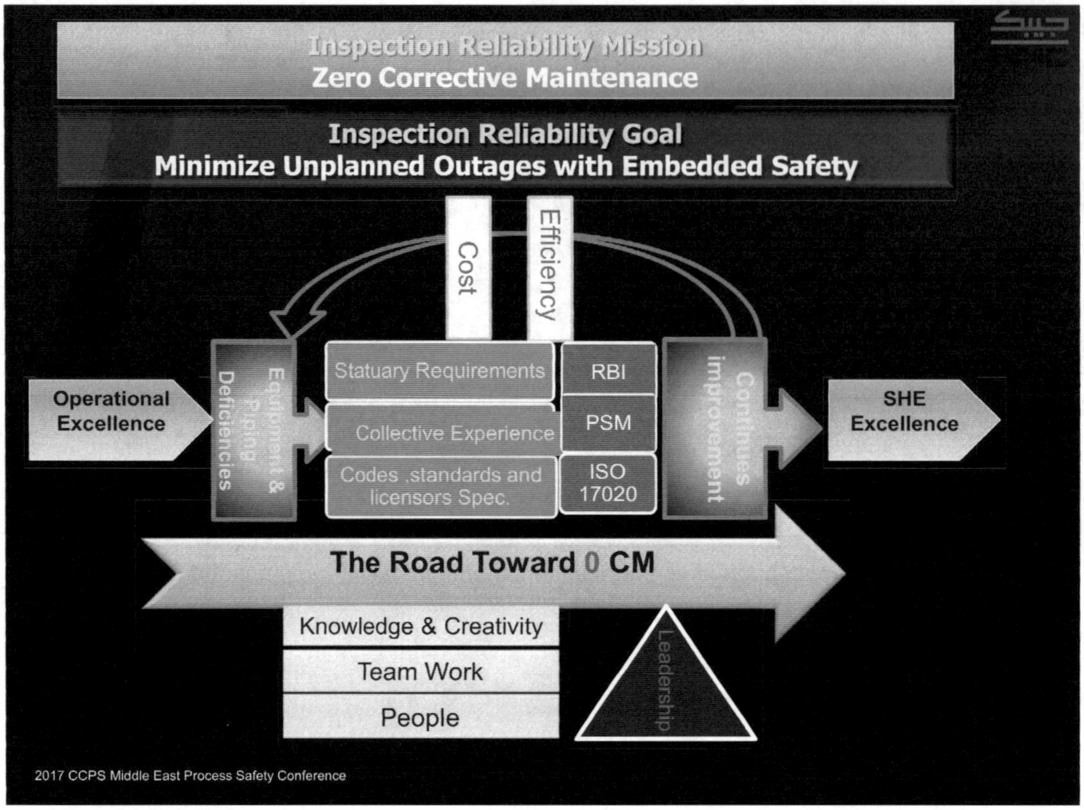
Inspection Reliability Mission
Zero Corrective Maintenance
Inspection Reliability Goal
Minimize Unplanned Outages with Embedded Safety
Cost
Efficiency
Operational Excellence
Equipment & Piping Deficiencies
Statuary Requirements
Collective Experience
Codes ,standards and licensors Spec.
RBI
PSM
ISO 17020
Continues improvement
SHE Excellence
The Road Toward 0 CM
Knowledge & Creativity
Team Work
People
Leadership
2017 CCPS Middle East Process Safety Conference

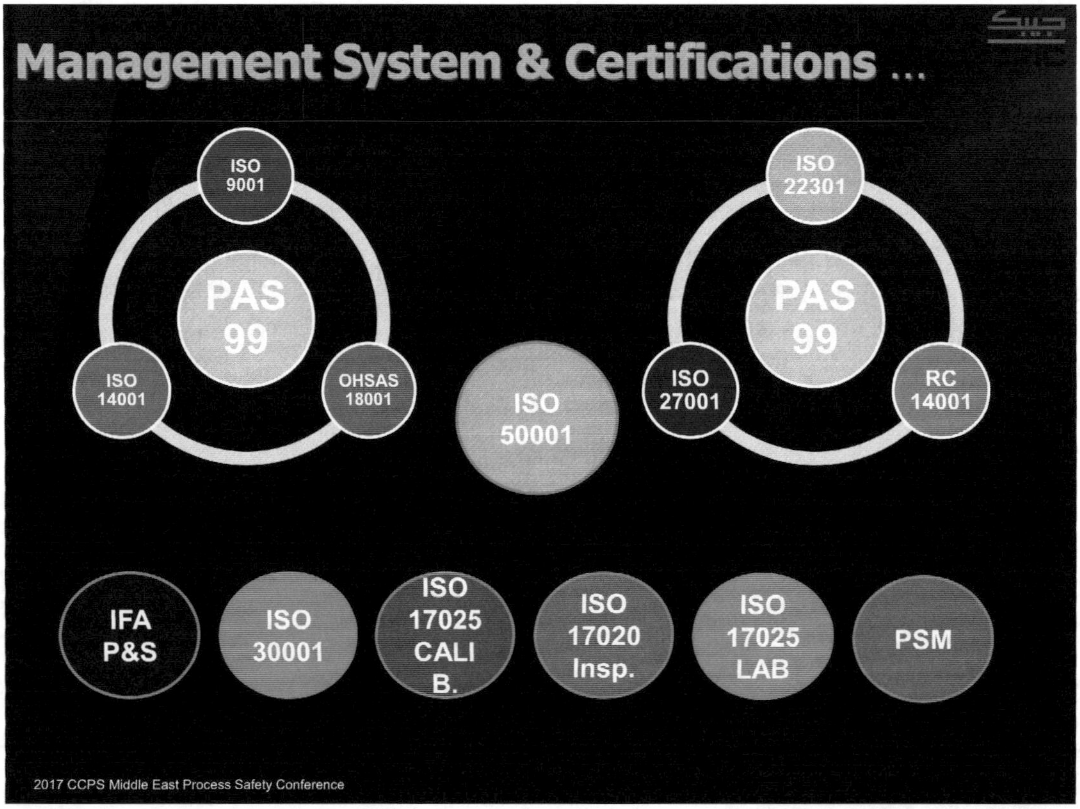
Management System & Certifications …
ISO 9001
PAS 99
ISO 14001
OHSAS 18001
ISO 50001
ISO 22301
PAS 99
ISO 27001
RC 14001
IFA P&S
ISO 30001
ISO 17025 CALI B.
ISO 17020 Insp.
ISO 17025 LAB
PSM
2017 CCPS Middle East Process Safety Conference

Integrated Reliability Systems

- Programs objective : is to Enhance the overall inspection practices & reliability with a structured, scientific & proven technology
Which is transparent & fully auditable

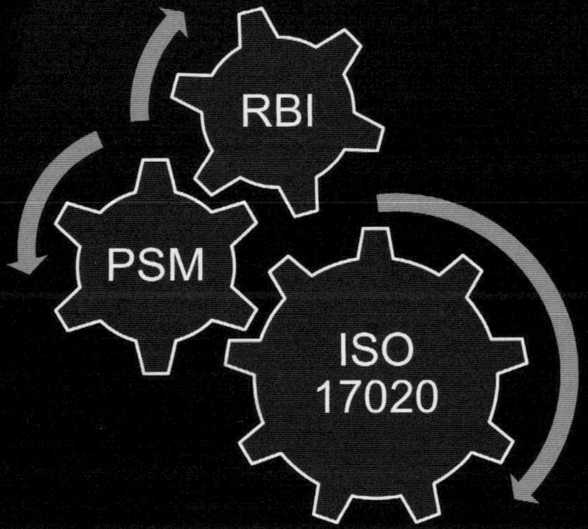

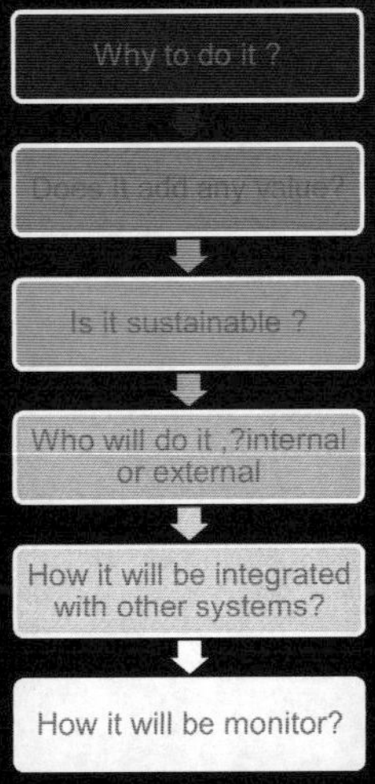

RBI Approach and integration

	Technical Services Department						
	Inspection Section						
RBI Completed							Remarks
	Ammonia	Methanol	Urea	Utilities	Product Handling & Offsite	Total	
Total (Planned)	116	152	104	58	10	440	
Total RBI Completed	100%	100%	100%	100%	100%	100%	

	Technical Services Department					Date: 26-09-2017
	Inspection Section					Rev. 00
Revalidation of RBI Study						Remarks
Year	Ammonia	Methanol	Urea	Utilities	Total	
Total Revalidations	131	165	108	90	494	includes 54 Piping Corrosion Loops

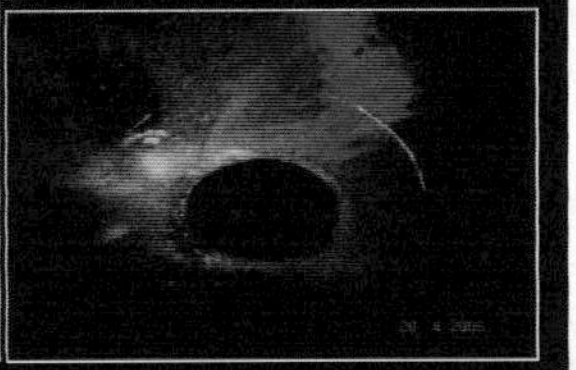

Around 20% enhancement in reliability assurance

PSM Approach and integration

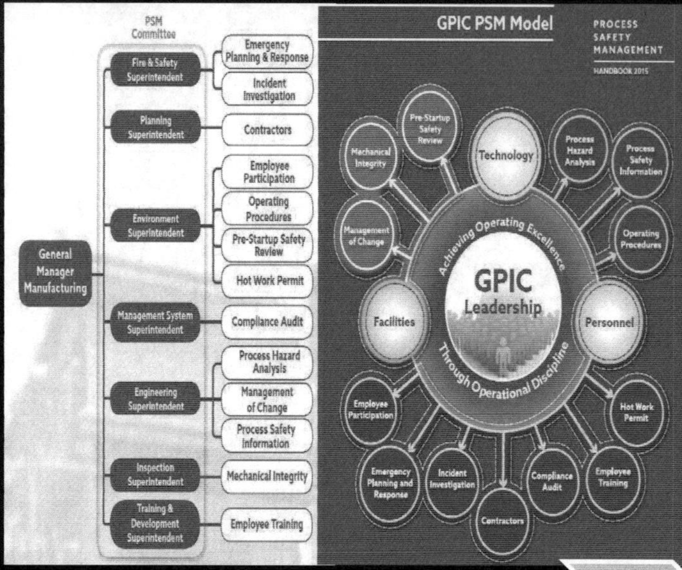

	PSM Elements	Key Performance Indicators
1	Employee Participation	A. Number of shift/safety talks completed B. Number of operators attending HAZOP studies
2	Process Safety Information	Number of outstanding documents update (P&IDs, MSDS, ...etc.)
3	Process Hazard Analysis	Number of outstanding PHA recommendations
4	Operating Procedures	Number of outstanding PSM related procedures pending review
5	Employee Training	Number of individuals with overdue refresher training
6	Contractors	Number of outstanding "contractor of the job audit" recommendations
7	Pre-Startup Safety Review	Number of outstanding PSSR
8	Mechanical Integrity	A. Number of overdue inspections B. The number of challenges to automated safety devices, such as number of trips
9	Hot Work Permit	Number of PTW violations
10	Management of Change	Number of outstanding MOCs
11	Incident Investigation	The number of outstanding PSM incident recommendations closed out
12	Emergency Planning and Response	Number of PSM emergency mock drills carried out
13	Compliance Audit	Number of PSM audits carried out

ISO 17020 Approach and integration

GPIC — General Management Inspection Quality Management System

GQS No. : IQMM
Issue No.: 1
Rev. No. : 3
Date : 25/08/2016
Page : 2 of 31

Inspection Quality Management Manual — Copy No.:

Table of contents

Where did it help?

- Integrated inspection manual
- Detail Checklists
- Performance Evaluation
- Knowledge Transfer
- Auditable System
- Assurance and structure

Inspection Reliability Programs

Emphasis on Preventive to Minimize Corrective Maintenance

1-Positive Material Identification	11-Equipment With Limited Access
2- Pre-Fabrication Inspection	12-Strategic Replacement -10 Yr
3-3rd party QA/QC	13-Root Cause Failure Analysis
4-Risk Based Inspection	14- Bi-Yearly Report Analysis
5- Equipment/Piping Condition Monitoring	On-line Condition monitoring
6-Underground & STA Survey	Applied Experience
7-Corrosion Under Insulation	Thermography Expanded
8-General Corrosion Survey	Nonmetallic QA/QC
9-Fitness For Service	Pre Mod. RBI
10-Thermographic Survey	Inspection Support Network

Reliability Best Practices

1. Spare Part Verification: Pre-T/A: Main machine critical spare parts physical inspection and clearance check
2. Third Party Inspection: Additional Quality controllers for stationary equipments
3. Vendor recommendations tracking system
4. Detailed Risk Assessments.

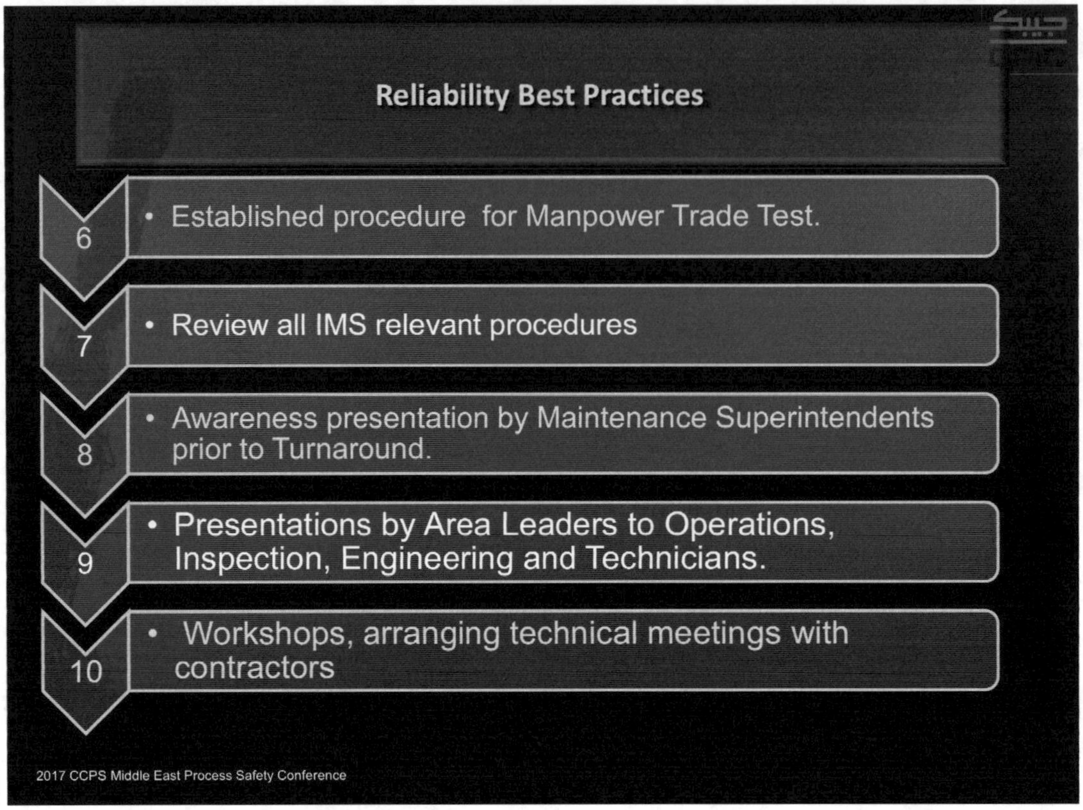
Reliability Best Practices
6
• Established procedure for Manpower Trade Test.
7
• Review all IMS relevant procedures
8
• Awareness presentation by Maintenance Superintendents prior to Turnaround.
9
• Presentations by Area Leaders to Operations, Inspection, Engineering and Technicians.
10
• Workshops, arranging technical meetings with contractors
2017 CCPS Middle East Process Safety Conference

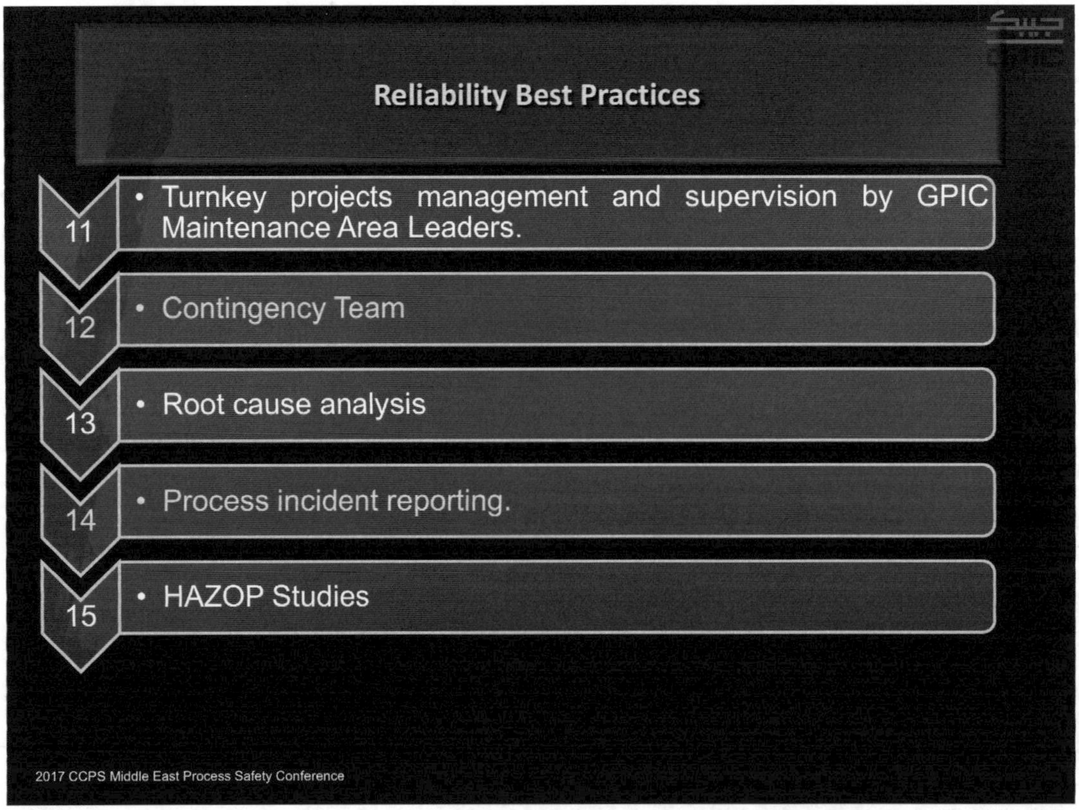
Reliability Best Practices
11
• Turnkey projects management and supervision by GPIC Maintenance Area Leaders.
12
• Contingency Team
13
• Root cause analysis
14
• Process incident reporting.
15
• HAZOP Studies
2017 CCPS Middle East Process Safety Conference

Reliability Best Practices

16 • Issuing Rolling 10 years Turnaround plan (2012 to 2022)

17 • TA Briefing & Debriefing

18 • Early appointment of Area Leaders

19 • Internal manpower utilization

20 • Classification of TA action reports

2017 CCPS Middle East Process Safety Conference

Reliability Best Practices

21 • Early mobilization of Resources

23 • Calibration, Testing and certification of Lifting and Measuring tools

24 • Acoustic Valve Leak Test.

25 • Modern Electrical Testing Techniques: Thermograph / Motor Circuit Evaluator

2017 CCPS Middle East Process Safety Conference

Reliability Best Practices

26 • E-learning

27 • Critical Joints Tightening: Dedicated team with specialist supervisor

28 • In-house house design, engineering and implementation

29 • Blinds Management

30 • Enhanced Night Shift Coverage

2017 CCPS Middle East Process Safety Conference

Thank You

2017 CCPS Middle East Process Safety Conference

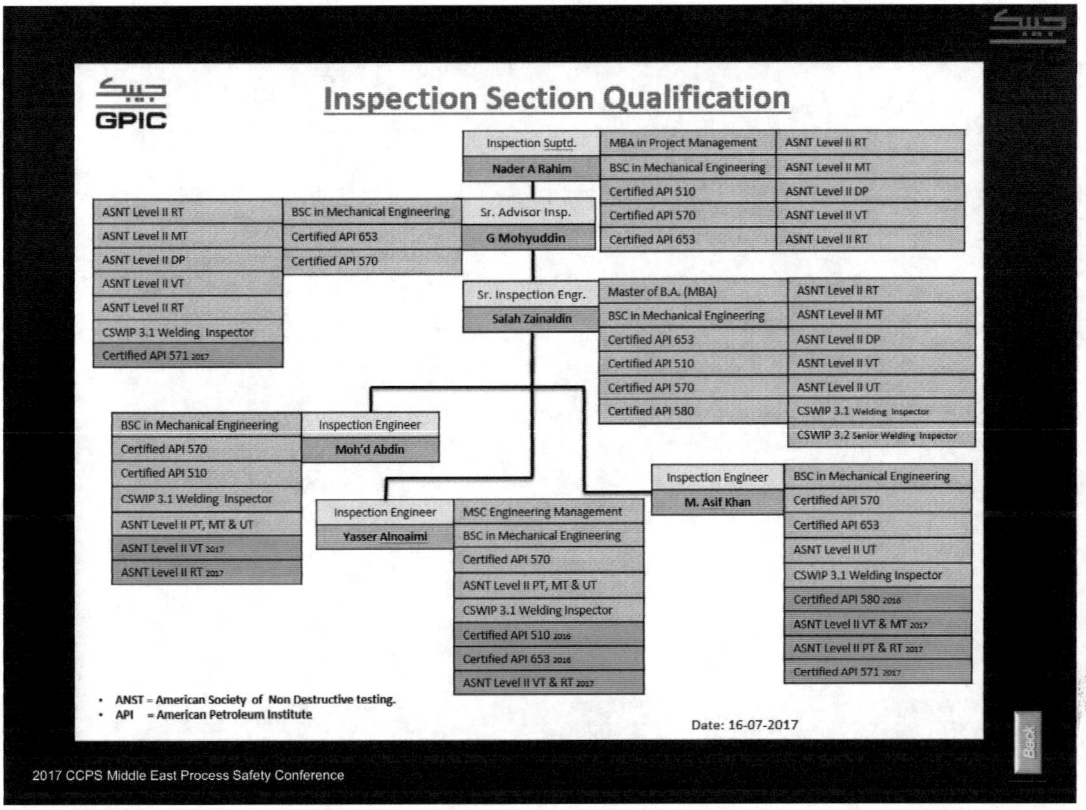
GPIC
Inspection Section Qualification
Inspection Suptd.
Nader A Rahim
MBA in Project Management
BSC in Mechanical Engineering
Certified API 510
Certified API 570
Certified API 653
ASNT Level II RT
ASNT Level II MT
ASNT Level II DP
ASNT Level II VT
ASNT Level II RT
Sr. Advisor Insp.
G Mohyuddin
BSC in Mechanical Engineering
Certified API 653
Certified API 570
ASNT Level II RT
ASNT Level II MT
ASNT Level II DP
ASNT Level II VT
ASNT Level II RT
CSWIP 3.1 Welding Inspector
Certified API 571 2017
Sr. Inspection Engr.
Salah Zainaldin
Master of B.A. (MBA)
BSC in Mechanical Engineering
Certified API 653
Certified API 510
Certified API 570
Certified API 580
ASNT Level II RT
ASNT Level II MT
ASNT Level II DP
ASNT Level II VT
ASNT Level II UT
CSWIP 3.1 Welding Inspector
CSWIP 3.2 Senior Welding Inspector
Inspection Engineer
Moh'd Abdin
BSC in Mechanical Engineering
Certified API 570
Certified API 510
CSWIP 3.1 Welding Inspector
ASNT Level II PT, MT & UT
ASNT Level II VT 2017
ASNT Level II RT 2017
Inspection Engineer
Yasser Alnoaimi
MSC Engineering Management
BSC in Mechanical Engineering
Certified API 570
ASNT Level II PT, MT & UT
CSWIP 3.1 Welding Inspector
Certified API 510 2016
Certified API 653 2016
ASNT Level II VT & RT 2017
Inspection Engineer
M. Asif Khan
BSC in Mechanical Engineering
Certified API 570
Certified API 653
ASNT Level II UT
CSWIP 3.1 Welding Inspector
Certified API 580 2016
ASNT Level II VT & MT 2017
ASNT Level II PT & RT 2017
Certified API 571 2017
• ANST = American Society of Non Destructive testing.
• API = American Petroleum Institute
Date: 16-07-2017
Back
2017 CCPS Middle East Process Safety Conference

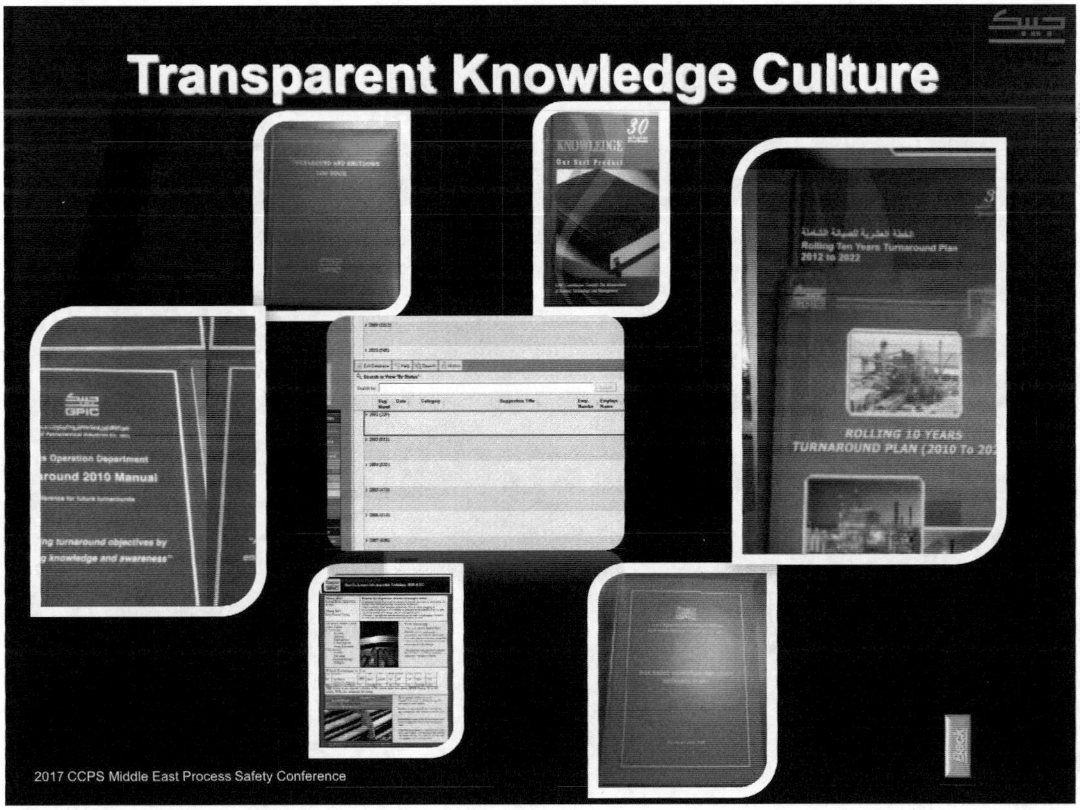
Transparent Knowledge Culture
KNOWLEDGE
Our Best Product
Rolling Ten Years Turnaround Plan
2012 to 2022
ROLLING 10 YEARS
GPIC
Back
2017 CCPS Middle East Process Safety Conference

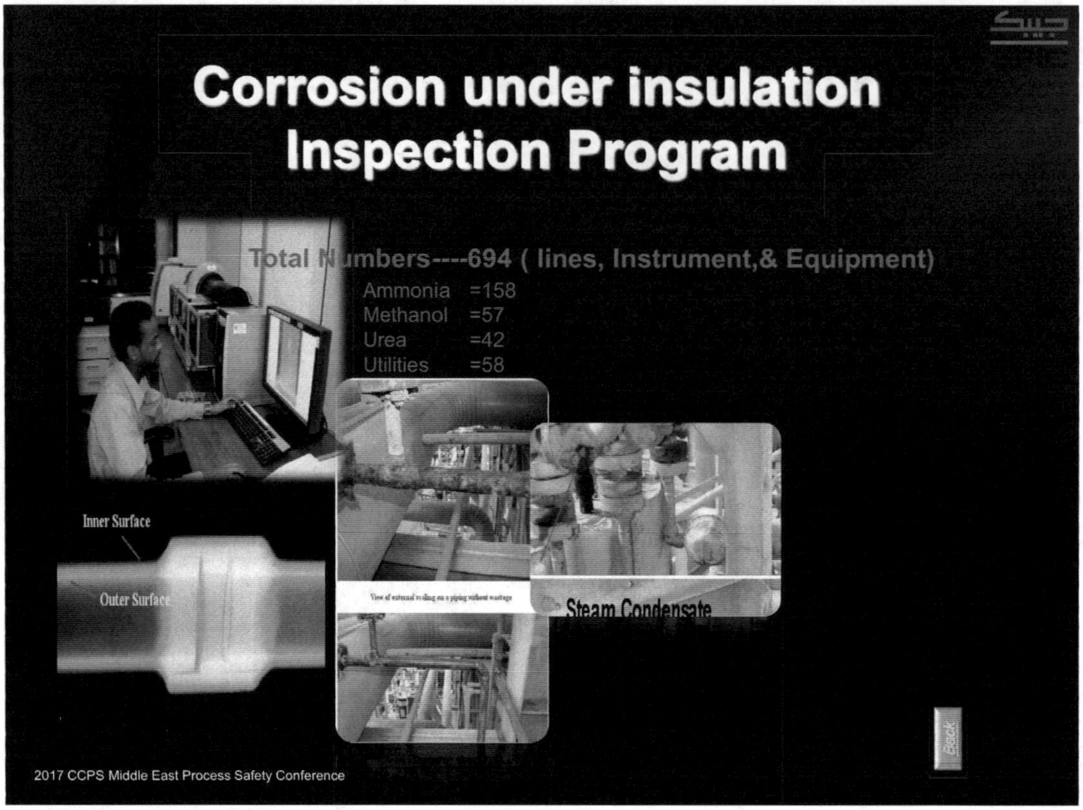
Corrosion under insulation Inspection Program
Total Numbers----694 (lines, Instrument,& Equipment)
Ammonia =158
Methanol =57
Urea =42
Utilities =58
Inner Surface
Outer Surface
Steam Condensate
Back
2017 CCPS Middle East Process Safety Conference

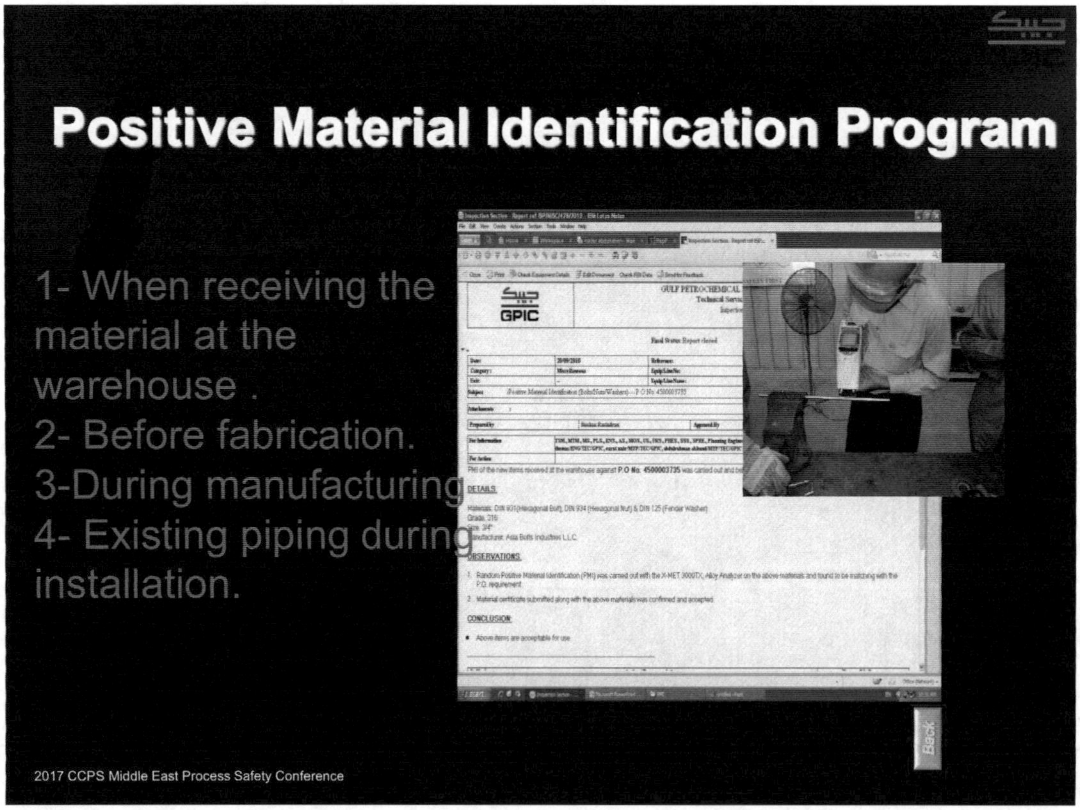
Positive Material Identification Program
1- When receiving the material at the warehouse .
2- Before fabrication.
3-During manufacturing
4- Existing piping during installation.
GPIC
Back
2017 CCPS Middle East Process Safety Conference

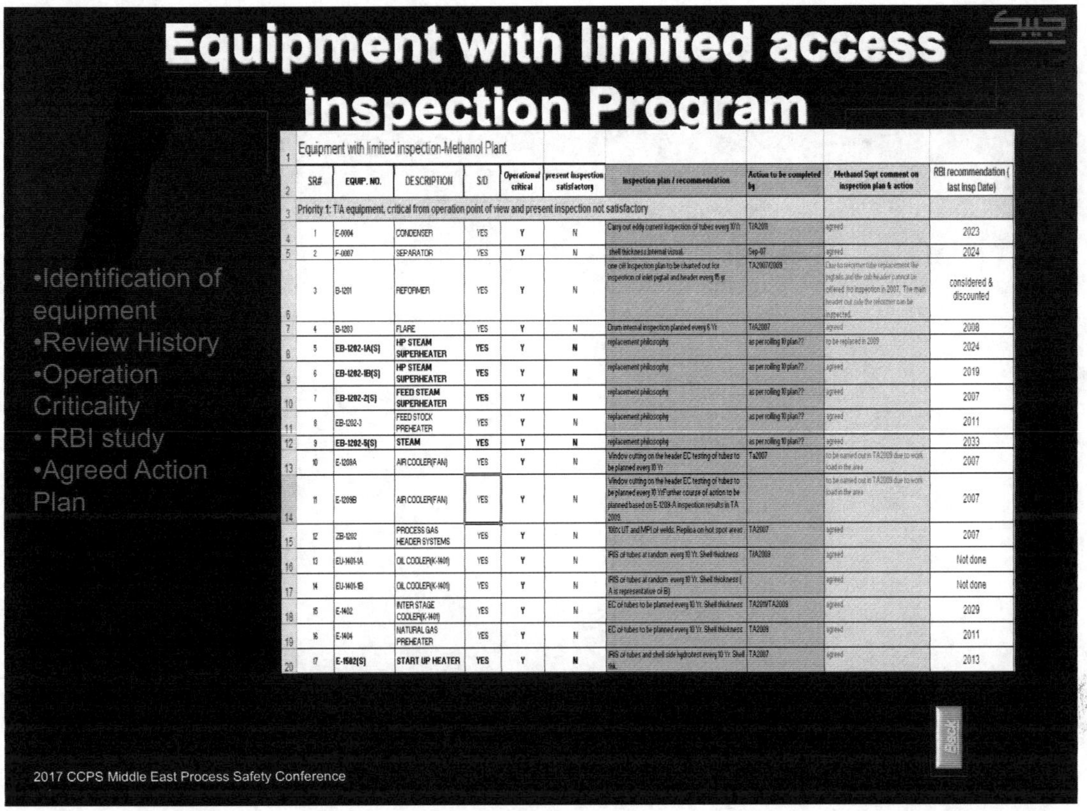

Equipment with limited inspection-Methanol Plant

SR#	EQUIP. NO.	DESCRIPTION	S/D	Operational critical	present inspection satisfactory	Inspection plan / recommendation	Action to be completed by	Methanol Supt comment on inspection plan & action	RBI recommendation (last insp Date)
Priority 1: T/A equipment, critical from operation point of view and present inspection not satisfactory									
1	E-0004	CONDENSER	YES	Y	N	Carry out eddy current inspection of tubes every 10 Yr	T/A2011	agreed	2023
2	F-0007	SEPARATOR	YES	Y	N	shell thickness internal visual	Sep-07	agreed	2024
3	B-1201	REFORMER	YES	Y	N	one off inspection plan to be charted out for inspection of inlet pigtail and header every 15 yr	TA2007/2009	Due to reformer tube replacement the pigtails and the sub header cannot be offered for inspection in 2007. The main header out side the reformer can be inspected.	considered & discounted
4	B-1203	FLARE	YES	Y	N	Drum internal inspection planned every 6 Yr	T/A2007	agreed	2008
5	**EB-1202-1A(S)**	**HP STEAM SUPERHEATER**	**YES**	Y	**N**	replacement philosophy	as per rolling 10 plan??	to be replaced in 2009	2024
6	**EB-1202-1B(S)**	**HP STEAM SUPERHEATER**	**YES**	Y	**N**	replacement philosophy	as per rolling 10 plan??	agreed	2019
7	**EB-1202-2(S)**	**FEED STEAM SUPERHEATER**	**YES**	Y	**N**	replacement philosophy	as per rolling 10 plan??	agreed	2007
8	EB-1202-3	FEED STOCK PREHEATER	YES	Y	N	replacement philosophy	as per rolling 10 plan??	agreed	2011
9	**EB-1202-5(S)**	**STEAM**	**YES**	Y	**N**	replacement philosophy	as per rolling 10 plan??	agreed	2033
10	E-1209A	AIR COOLER(FAN)	YES	Y	N	Window cutting on the header EC testing of tubes to be planned every 10 Yr	Ta2007	to be carried out in TA2009 due to work load in the area	2007
11	E-1209B	AIR COOLER(FAN)	YES	Y	N	Window cutting on the header EC testing of tubes to be planned every 10 YrFurther course of action to be planned based on E-1209-A inspection results in TA 2009		to be carried out in TA2009 due to work load in the area	2007
12	ZB-1202	PROCESS GAS HEADER SYSTEMS	YES	Y	N	100% UT and MPI of welds. Replica on hot spot areas	TA2007	agreed	2007
13	EU-1401-1A	OIL COOLER(K-1401)	YES	Y	N	IRIS of tubes at random every 10 Yr. Shell thickness	T/A2009	agreed	Not done
14	EU-1401-1B	OIL COOLER(K-1401)	YES	Y	N	IRIS of tubes at random every 10 Yr. Shell thickness (A is representative of B)		agreed	Not done
15	E-1402	INTER STAGE COOLER(K-1401)	YES	Y	N	EC of tubes to be planned every 10 Yr. Shell thickness	TA2011/TA2009	agreed	2029
16	E-1404	NATURAL GAS PREHEATER	YES	Y	N	EC of tubes to be planned every 10 Yr. Shell thickness	TA2009	agreed	2011
17	**E-1502(S)**	**START UP HEATER**	**YES**	Y	**N**	IRIS of tubes and shell side hydrotest every 10 Yr. Shell thk.	TA2007	agreed	2013

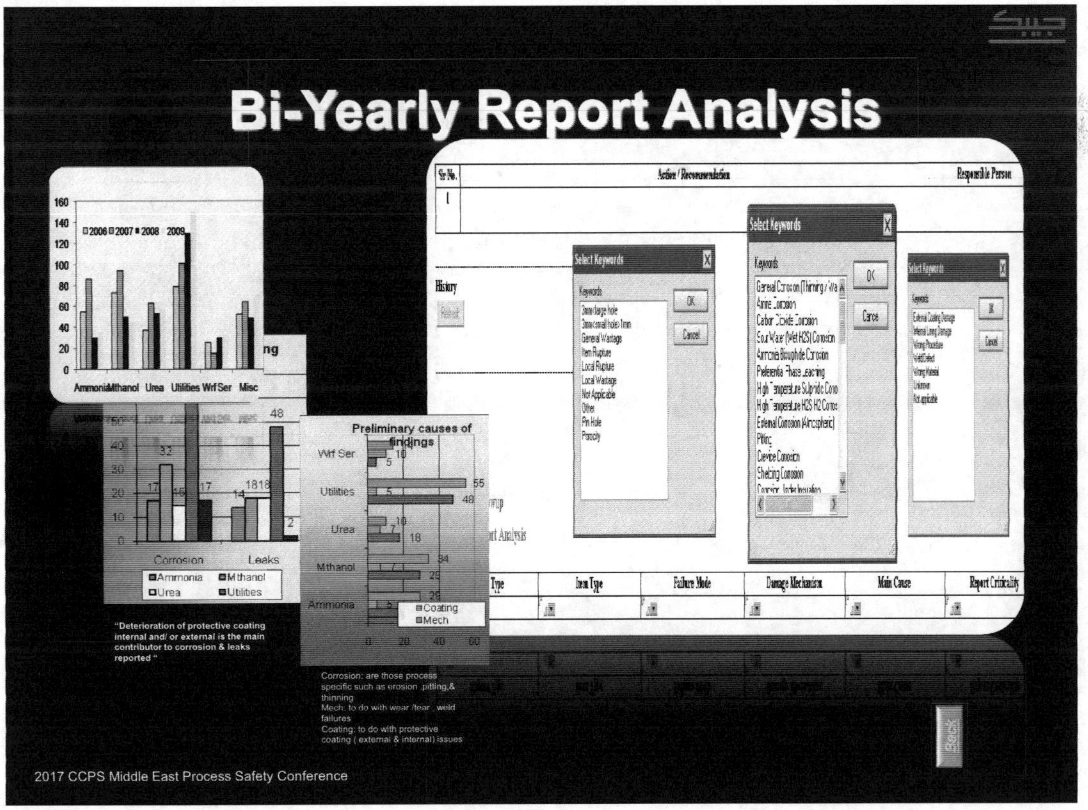

Root Cause Failure Analysis Program

- TapRoot Methodology
- 35 trained in TapRoot
- RCFA Database
- Dedicated SOP

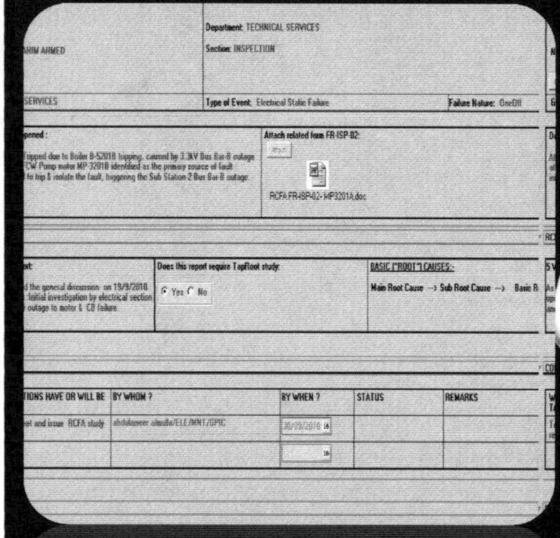

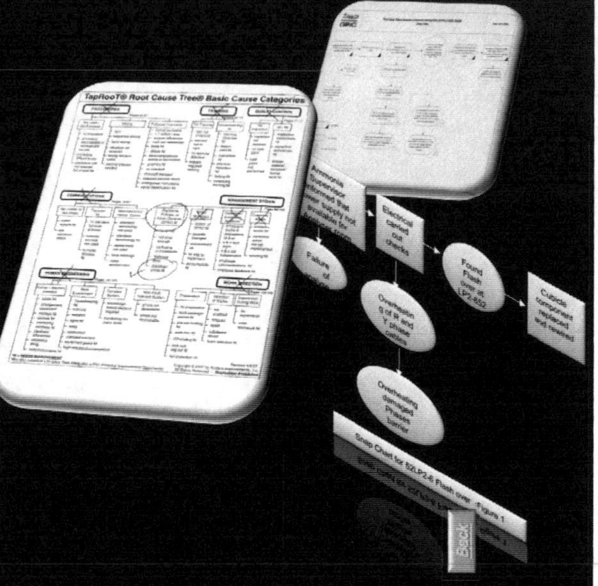

Fitness For Service Program

- Specialized Service provider
- FFS on operating & removed items.
- Confirmation process.
- Regular intervals

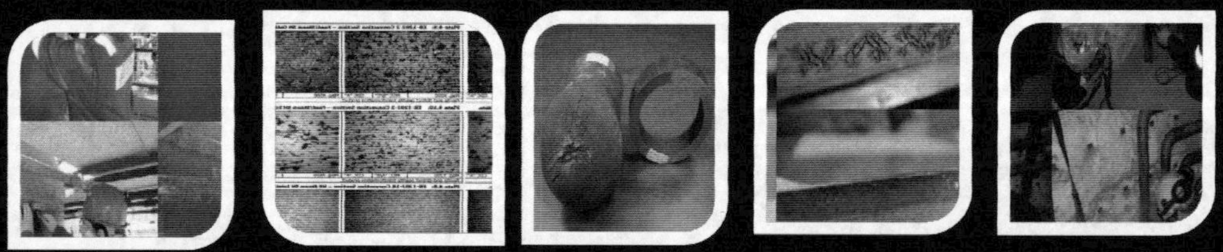

Back

2017 CCPS Middle East Process Safety Conference

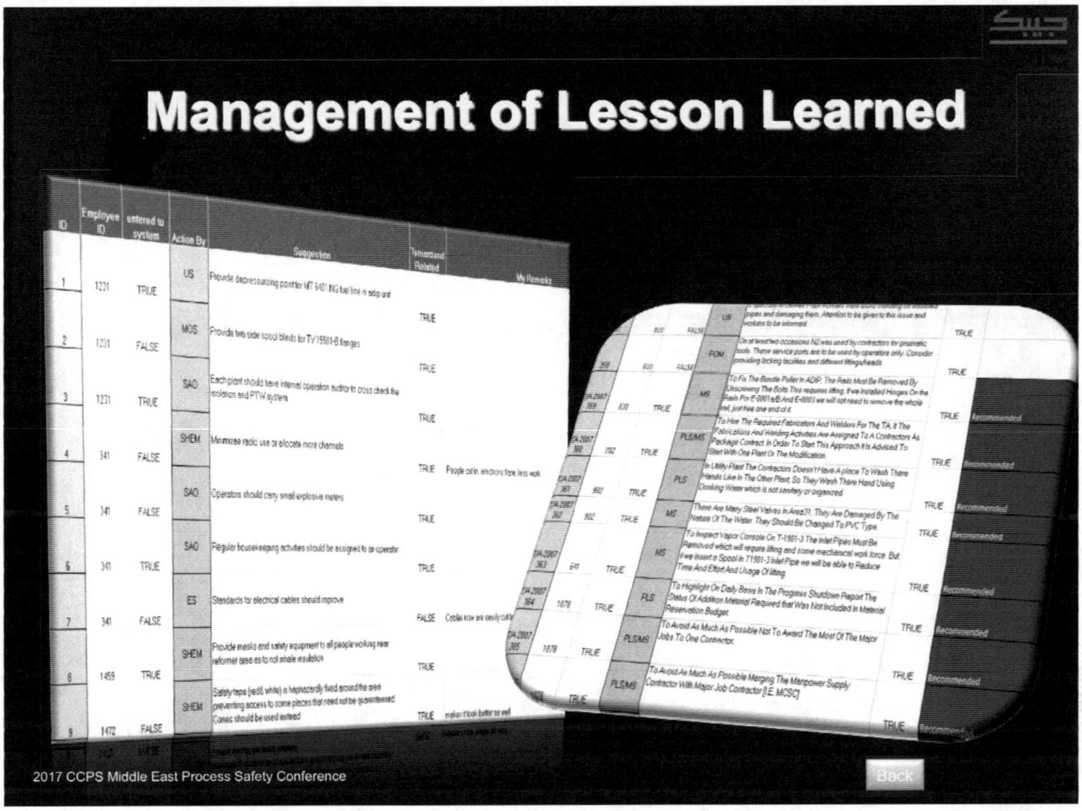
Management of Lesson Learned
2017 CCPS Middle East Process Safety Conference
Back

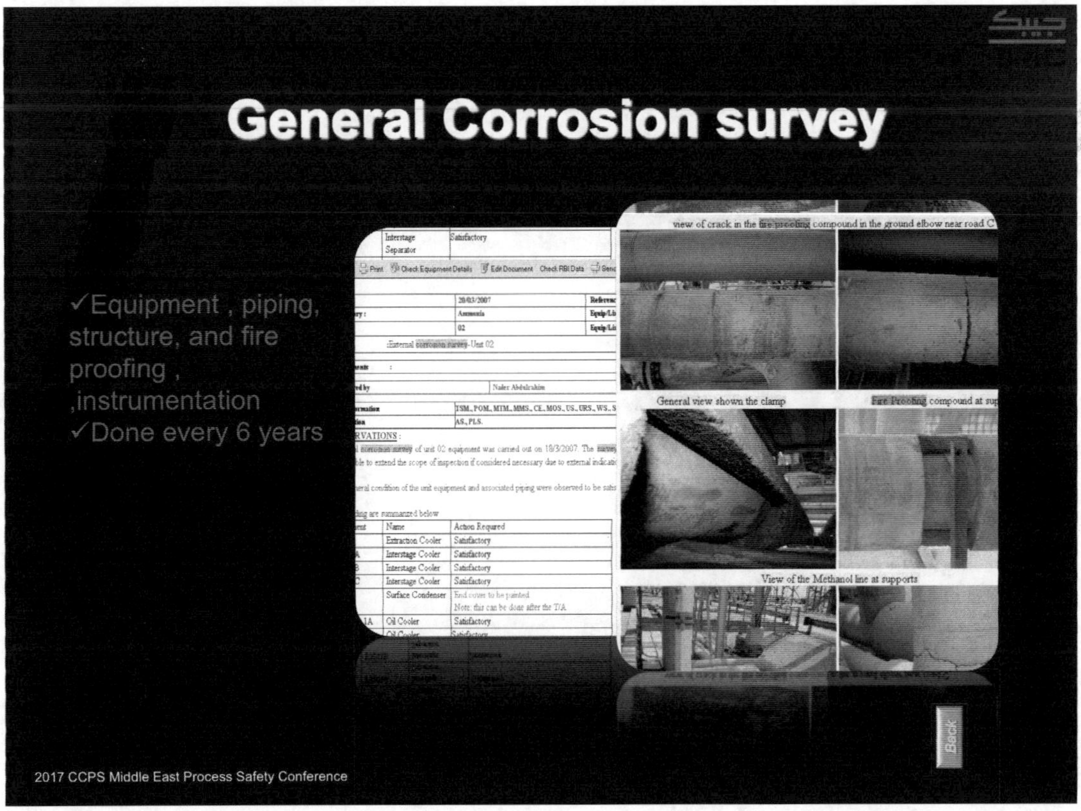
General Corrosion survey
✓Equipment , piping, structure, and fire proofing , ,instrumentation
✓Done every 6 years
2017 CCPS Middle East Process Safety Conference
Back

Communicating Hazards in a Multicultural Workforce

CCPS Middle East Process Safety Conference
11 October 2017, Bahrain

Atif Mohammed Ashraf and Dr. Luc Vechot
MKOPSC at TAMU

Dr. Stephanie C. Payne and Dr. Nathanael L. Keiser
Department of Psychology and MKOPSC at TAMU

1

Overview

- Mary Kay O' Connor Process Safety Center
- Background
- Methodology
- Results and Analysis

2

Texas A&M University College of Engineering

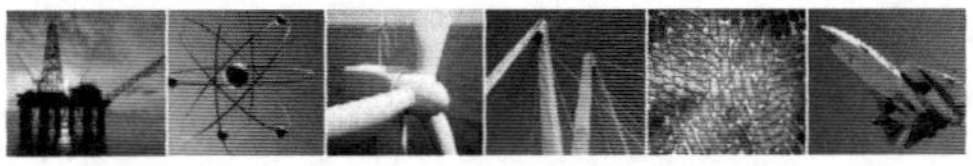

- Part of Texas A&M University since its inception in 1876
- 14 departments.
- 500 faculty members
- 16,000 engineering students

3

Mary Kay O'Connor Process Safety Center

Making Safety Second Nature

To be an international leader in minimizing losses within the process industry – through safer processes, management, equipment, and procedures.

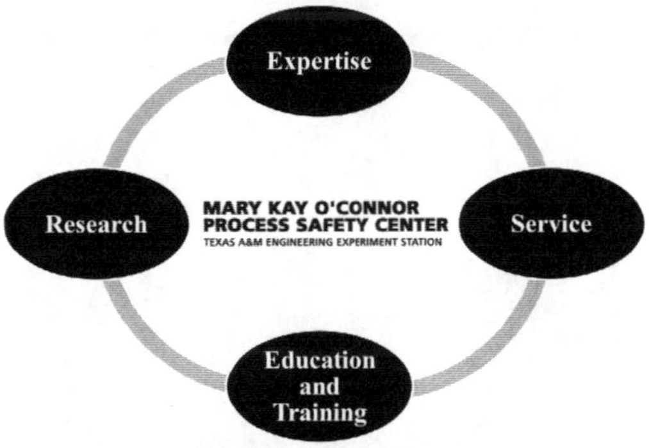

Center established in 1995 in memory of Mary Kay O'Connor (Chemical engineer and Operations Superintendent) who died in the 1989 Chemical plant accident- Pasadena, TX

4

Mary Kay O'Connor Process Safety Center

- Provide a national and regional **hub** of Process Safety expertise
- Provide the industry in the Middle East a **forum** for exchange of ideas, best practices, advances in proves safety
- Conduct **research** on process safety topic relevant to the region
- **Educate** a new generation of engineers in the region who have process safety as second nature
- **Train** personnel
  - Regulatory institutions
  - Industry
  - Universities

TEXAS A&M UNIVERSITY | MKOPSC MARY KAY O'CONNOR PROCESS SAFETY CENTER | LIBERAL ARTS TEXAS A&M UNIVERSITY PSYCHOLOGY

5

TEXAS A&M UNIVERSITY | MKOPSC MARY KAY O'CONNOR PROCESS SAFETY CENTER | LIBERAL ARTS TEXAS A&M UNIVERSITY PSYCHOLOGY

BACKGROUND

6

Mary Kay O'Connor Process Safety Center

"Toward Better Safety Communication in a Multicultural Workforce"

7

Workforce Challenges

- International corporations produce goods or services outside their home country
  - People from various nationalities and background work together
- Multinational workforces can experience communication challenges due to language barriers and cultural differences
- **National culture** influences communication and workplace safety
- **Communication** is critical to both personal and process safety

8

National Culture

- The collective programming of the mind that distinguishes the members of one group or category of people from others (Hofstede, 2001)
- "Software of the Mind"

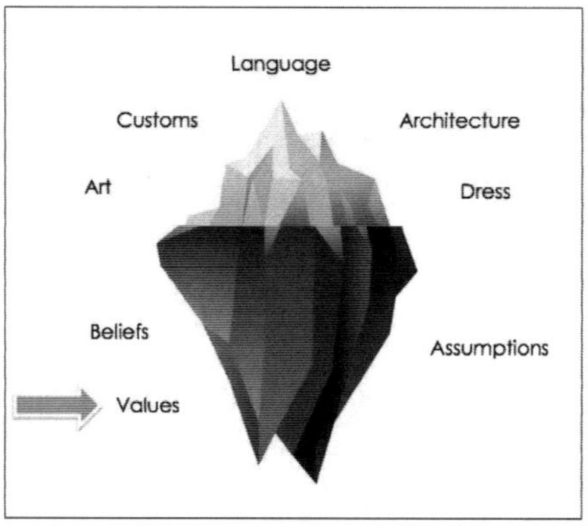

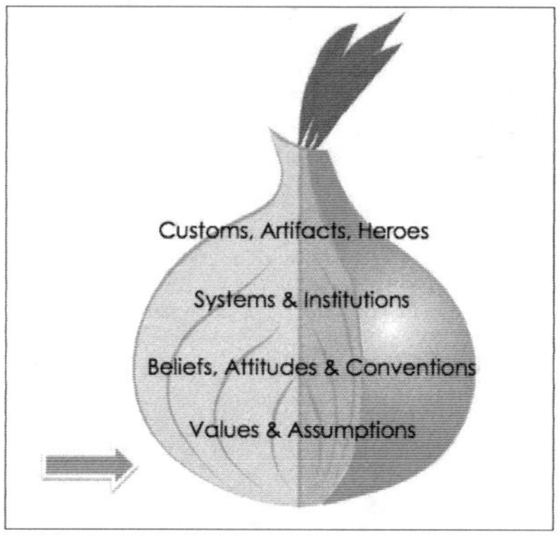

9

National Culture Dimensions

Individualism-Collectivism
Power Distance
Uncertainty Avoidance
Assertiveness
Long-term Orientation

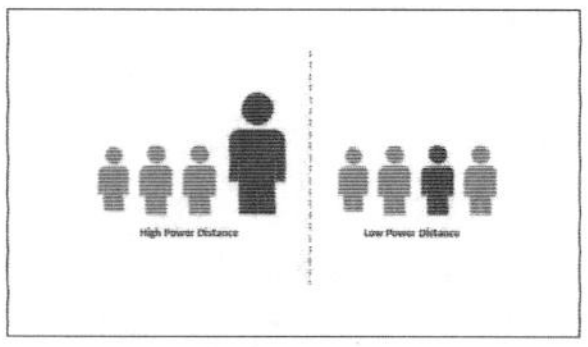

10

Safety Communication Challenge

- An example of communication challenge one face in working with individuals from different cultures/nationalities

- Employees are apprehensive to point out an unsafe condition or behavior to a supervisor
  - There are numerous examples of co-pilots not pointing out a known concern to pilots in the aviation industry. This has been effectively addressed with crew resource management (CREW) training (Helmreich, Merritt, & Wilhelm, 1999)

Power Distance

How comfortable people are with speaking to authority figures

11

METHODOLOGY

12

Methodology

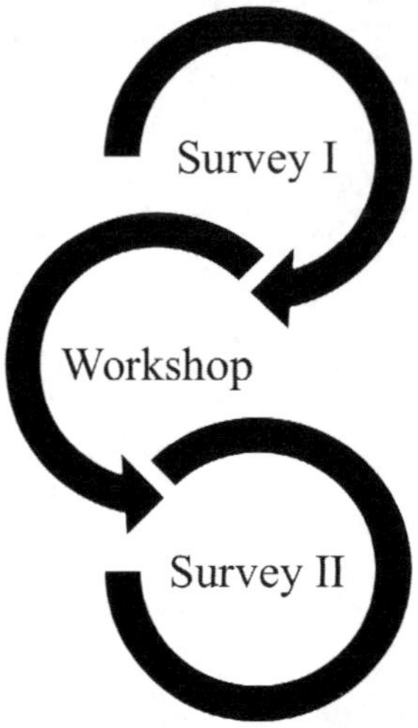

Administer pre-workshop survey to gain insights about current challenges at Company

Develop safety communication workshop tailored to the Company's needs

Conduct the workshop with a subset of the Company's workforce

Administer post-workshop survey to determine efficacy of the workshop

13

Survey Objectives

- Gain insights into safety-related communication challenges at the company
  - Gather information directly from employees
  - Identify strengths/areas for improvement
- Use the results in the development/assessment of a communication workshop
  - Provide baseline information about employees' standing on survey constructs
- Share results with company employees and management
  - Provide research-based, practical recommendations

14

Survey Constructs

- **Communication-Related Constructs**
  - Quality of Safety Communication
  - Safety Communication Encouragement
  - Clarity of Safety-Related Communication Comprehension
  - Global Communication Effectiveness
- **National Culture Dimensions**
  - Collectivism
  - Power Distance
  - Assertiveness
  - Uncertainty Avoidance
  - Long-term Orientation
- **Safety-Related Psychological Constructs**
  - Risk Propensity
  - Perceived Harm
  - Safety Climate
  - Safety Knowledge
  - Safety Motivation
  - Safety Compliance
  - Safety Participation
- **Self-Reported Injuries, Incidents, Near Misses (last 6 months)**

15

RESULTS AND ANALYSIS: SURVEY 1

16

Sample Demographics

TEXAS A&M UNIVERSITY | TEES MKOPSC | LIBERAL ARTS TEXAS A&M UNIVERSITY PSYCHOLOGY

Department

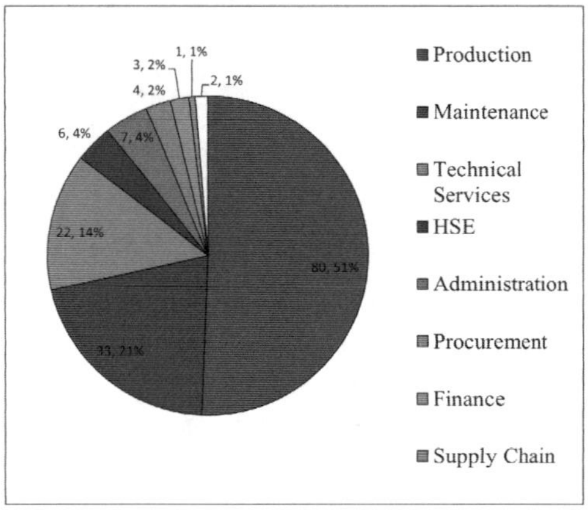

Level

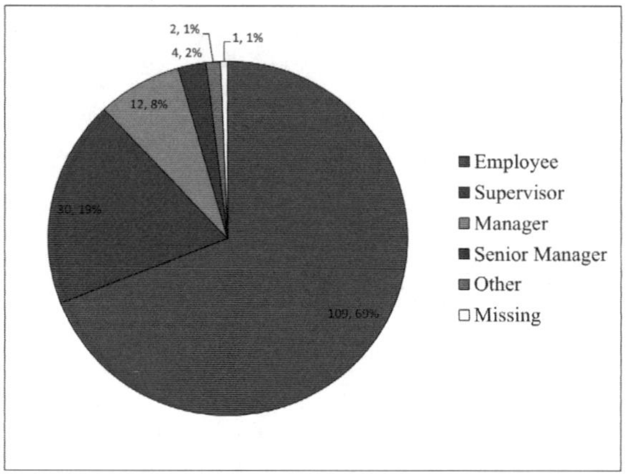

17

Nationality

TEXAS A&M UNIVERSITY | TEES MKOPSC | LIBERAL ARTS TEXAS A&M UNIVERSITY PSYCHOLOGY

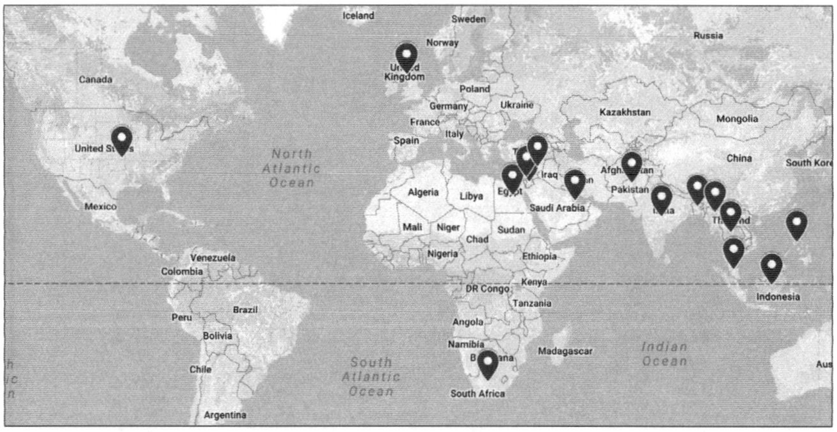

- **Bangladesh (6)**
- **Burma (1)**
- **Egypt (3)**
- **India (166)**
- **Indonesia (7)**
- **Jordan (8)**
- **Malaysia (6)**
- **Pakistan (60)**
- **Palestine (2)**
- **Philippines (6)**
- **Qatar (35)**
- **South Africa (1)**
- **Syria (1)**
- **Thailand (1)**
- **United Kingdom (2)**
- **United States (1)**

18

Clarity of Safety-Related Communication

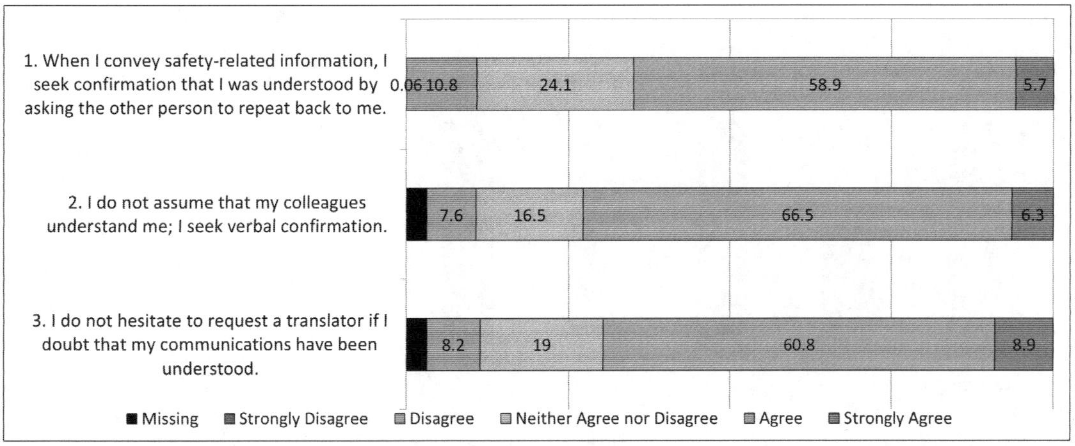

Workplace Safety Communication

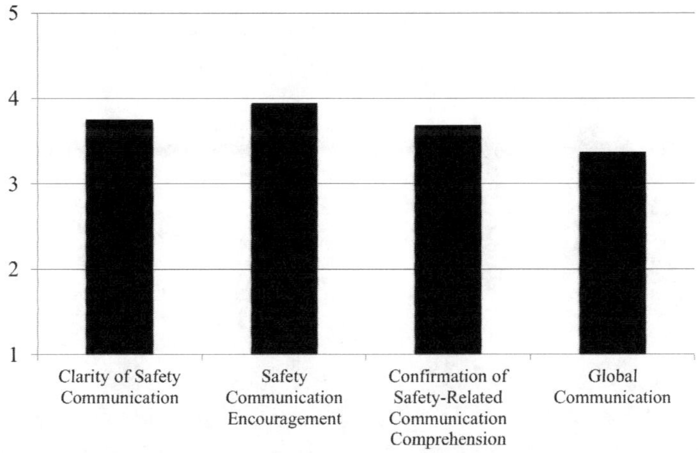

1. **Clarity of Safety Communication** – Timeliness, accuracy, usefulness, and quantity of safety-related information
2. **Safety Communication Encouragement** – Encouraged and feel comfortable engaging in safety-related communications
3. **Confirmation of Safety-Related Communication Comprehension** – Ensure that safety communication is understood
4. **Global Communication Effectiveness** – Effectiveness of practices, procedures, and policies regarding cross-cultural communication

Cultural Dimensions

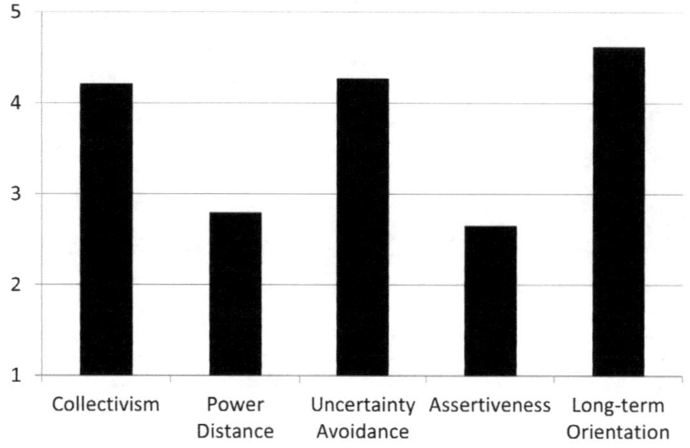

1. **Collectivism** – Interdependent vs. independent
2. **Power Distance** – Acceptance of social hierarchy
3. **Uncertainty Avoidance** – Follow strict rules and guidelines for behavior
4. **Assertiveness** – Dominance vs. social considerations
5. **Long-term Orientation** – Future rewards vs. respect for tradition

21

Safety Constructs

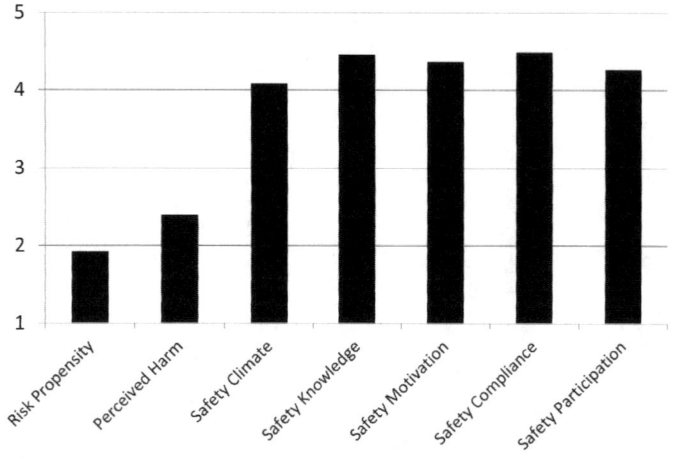

1. **Risk propensity** – Inclined to take risks, seek adventure, and engage in risky behaviors
2. **Perceived Harm** – Extent of job risk and beliefs about the effectiveness of controls
3. **Safety Climate** –The priority of safety
4. **Safety Knowledge** – Understanding of how to work safely
5. **Safety Motivation** – Willingness to exert effort to enact safety behaviors
6. **Safety Compliance** – Core activities to maintain workplace safety
7. **Safety Participation** – Efforts that support and improves workplace safety

22

Comparison – Level of Employee

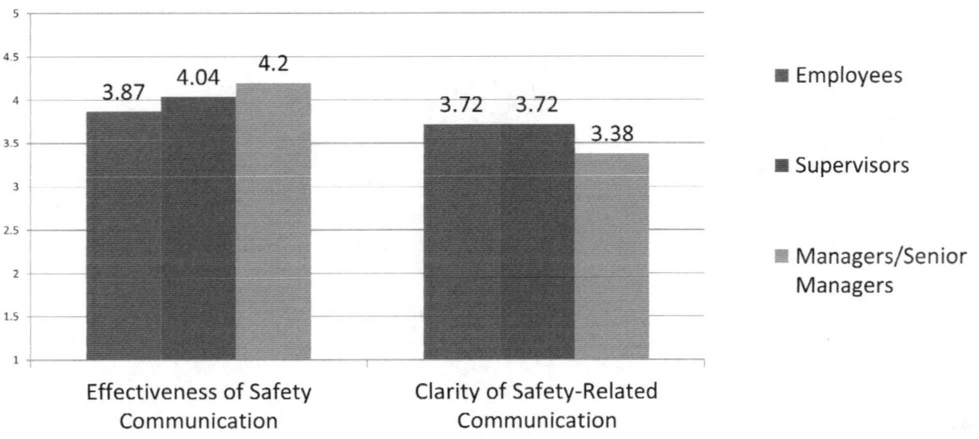

* $p < .05$

23

Comparison – Nationality

	India (N = 67)	Pakistan (N = 36)	Qatar (N = 11)	Other (N = 41)
Effectiveness of Safety Communication	4.05[a]	3.85[a]	3.98	3.84
Risk Propensity	1.89[a]	1.91	2.70[a,b]	1.82[b]
Perceived Harm	2.20[a,b]	2.51[a]	2.37	2.58[b]
Safety Knowledge	4.49[a]	4.49[b]	4.11[a,b]	4.43
Safety Motivation	4.43[a]	4.41	4.04[a]	4.28

Superscripts indicate significant differences $p < .05$

24

Comparison – Nationality Match

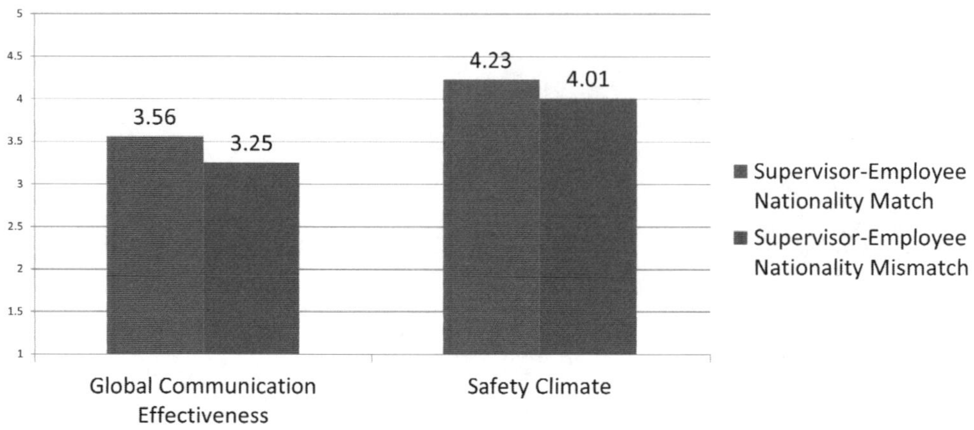

* $p < .05$

Open-ended Survey Comments

Summary/Recommendations based on Survey 1

1. **Promote safety communication**
   - Ensure that all employees feel comfortable speaking up about any safety-related concerns and reporting unsafe conditions, behaviors, and incidents without retribution
2. **Maintain and communicate a commitment to safety**
   - Continue to value and prioritize safety and communicate that priority consistently to all employees through resource allocation and never compromising safety for production
3. **Reward safe behavior**
   - Encourage supervisors to regularly acknowledge employees' safe behaviors and publicize these to the workgroup
4. **Offer safety training opportunities**
   - Monitor safety training needs and offer training and development opportunities when knowledge and skill deficiencies emerge

Work in Progress

- National culture – Safety Communication Workshop Delivered in July of 2017
- Survey 2 Administered in October of 2017

THANK YOU

29

aramco

Sustainable Corrective Actions - Is SMART Good Enough?

Upstream Operational Compliance Division

where energy is opportunity™

Current Corrective Action Methodologies & Deficiencies

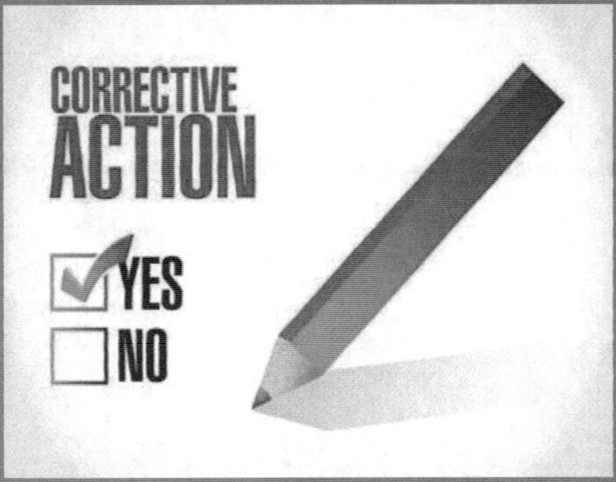

SMART

- First Use in journal, "Management Review" in 1982 by Doran
- Several different versions
- Much of the intent is very similar if not exactly the same
- Expanded to SMARTER©

S	**Specific** A defined end point or target of the web interaction
M	**Measurable** Attach numbers and timelines to the goal
A	**Achievable** Based on existing figures and research is it achievable?
R	**Relevant** Is this web goal in line with the business's broader goals?
T	**Time Limited** The goal must be measurable over a period of time
E	**Effective** The action must eliminate the associated root cause
R	**Reviewed** The action has been independently reviewed for unintended consequences

aramco

Field Review Results - Quality Issues

- Causal factor (CF) - A "problem" in the form of who did what wrong or what equipment failed or did not operate as intended
- Multiple problems (CFs) usually lead to the event
- Every CF has its own group of root causes
- There are one or more CARs to address each CF
- Wrong or missed problems; wrong root causes; wrong or weak CARs

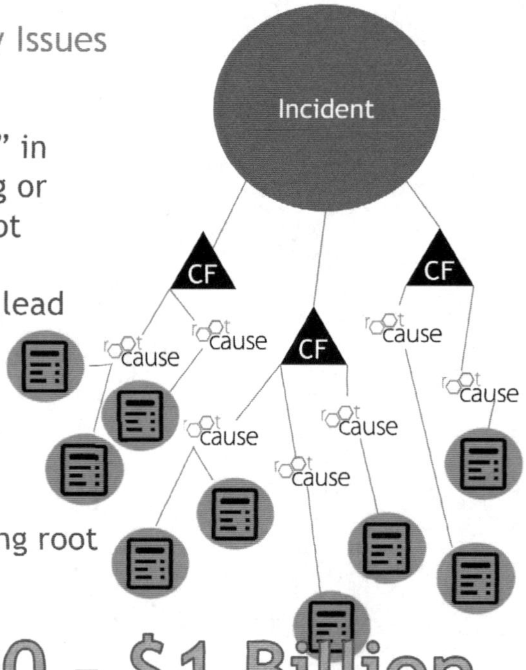

$750,000,000 - $1 Billion

aramco

Saudi Aramco: Public

Field Review Results - CARs

- Corrective Action Recommendations (CARs) not defined to address reason for root cause
- CARs don't address generic problems
- CARs often implemented to "check the box"
- "Measurement" most often addresses only the short-term
- CARs are not structured to address long-term sustainability
- CARs often cease to exist within a few months
- Incidents occur with repeat root causes/causal factors

Four Tools for Improving CAR Development & Implementation

Tool 1 – Quality Assurance Hold Points

- Problem identification (CF/SI selection)
- Root cause selection
- Suggested CARs
- Final report

Tool 2 - Connecting the Dots

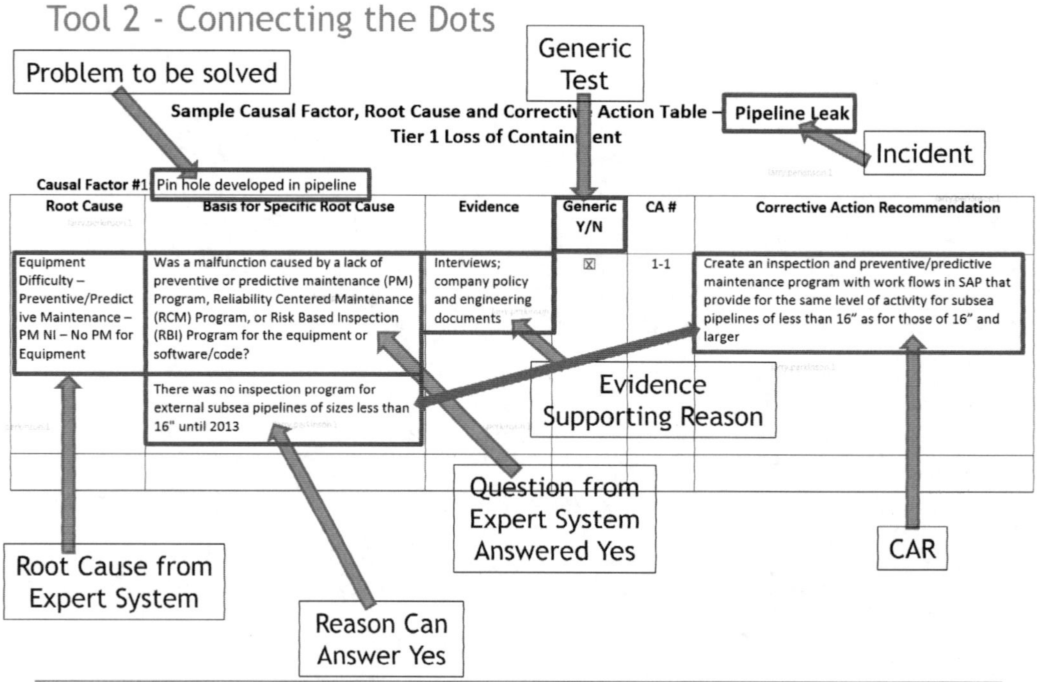

Sample Causal Factor, Root Cause and Corrective Action Table – Pipeline Leak
Tier 1 Loss of Containment

Causal Factor #1 Pin hole developed in pipeline

Root Cause	Basis for Specific Root Cause	Evidence	Generic Y/N	CA #	Corrective Action Recommendation
Equipment Difficulty – Preventive/Predictive Maintenance – PM NI – No PM for Equipment	Was a malfunction caused by a lack of preventive or predictive maintenance (PM) Program, Reliability Centered Maintenance (RCM) Program, or Risk Based Inspection (RBI) Program for the equipment or software/code?	Interviews; company policy and engineering documents	☒	1-1	Create an inspection and preventive/predictive maintenance program with work flows in SAP that provide for the same level of activity for subsea pipelines of less than 16" as for those of 16" and larger
	There was no inspection program for external subsea pipelines of sizes less than 16" until 2013				

Tool 3 - Building in Sustainability

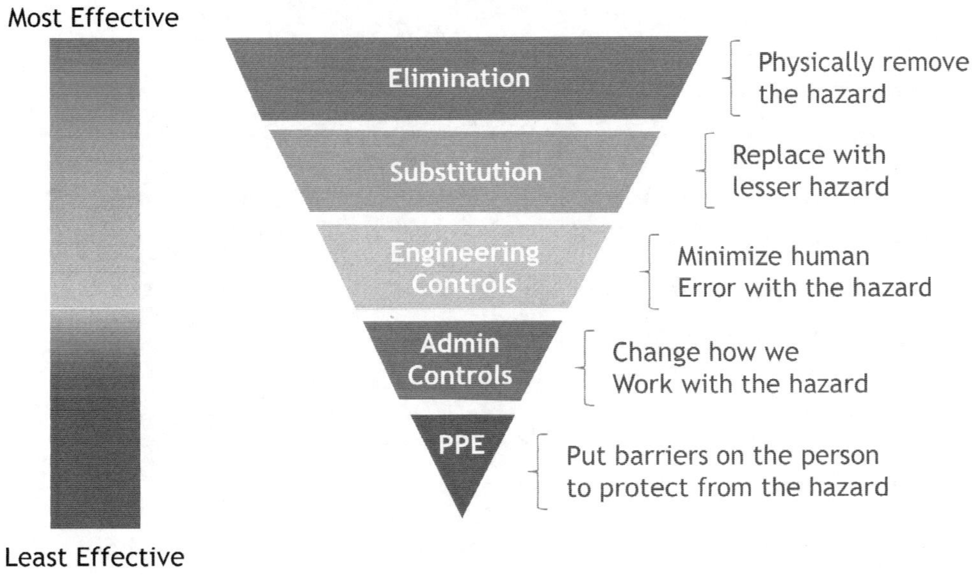

9

Saudi Aramco: Public

Tool 3 - Building in Sustainability

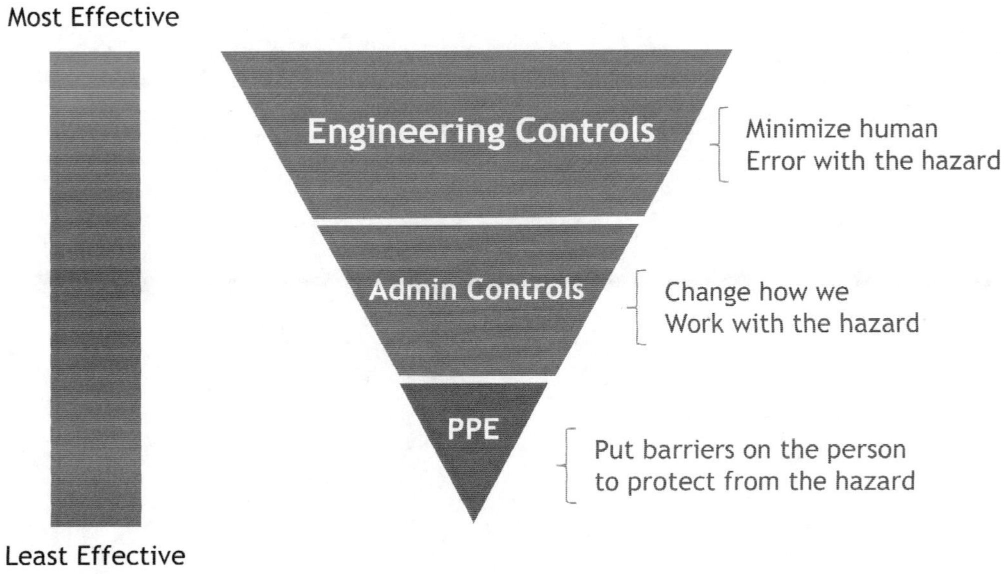

10

Saudi Aramco: Public

Tool 3 – Building in Sustainability – Engineering Controls

- Operates independently of
- human action
- Requires human intervention
- or equipment is sure to fail

- Inspection for control
- Predictive/Preventive
- Maintenance for control
- Training for maintenance personnel on control
- Competence Evaluations
- PM Data Gathering & Analysis
- Root Cause Analysis of equipment failures
- Fail Safe/Monitored
- Third Party Audits

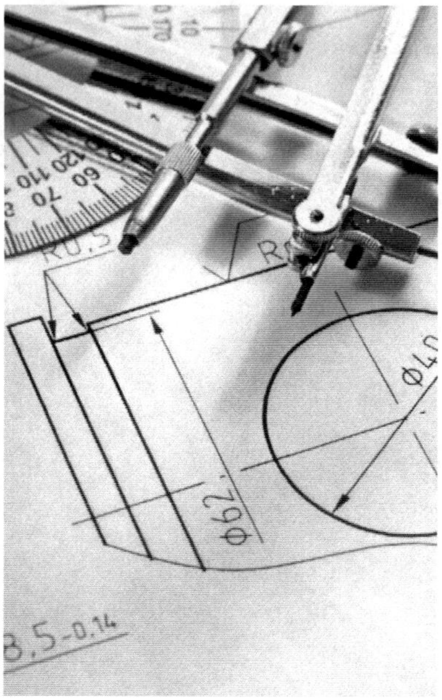

Tool 3 – Building in Sustainability – Admin Controls

- Training in the control
- Competency evaluation
- Provisions for new personnel
- Audits of training compliance
- Control in line with BP
- Audits of control application
- Reporting audit results to management
- Results used in KPI/PM
- Audit results analyzed for root causes and actioned
- MOC
- BBS
- Third party audits

Tool 3 – Building in Sustainability – PPE Controls

- Is suited to protect from the / Audit tied to disciplinary system
- Hazard / Training & competence
- Is designed to protect from the / PPE inspection & maintenance
- Energy level expected / Audits of PPE compliance
- Audits reported to management
- PPE selection & use in line with BP
- Audit results used in KPI/PM
- Audit results analyzed for root causes and actioned
- MOC
- PPE included in BBS
- Third party audit

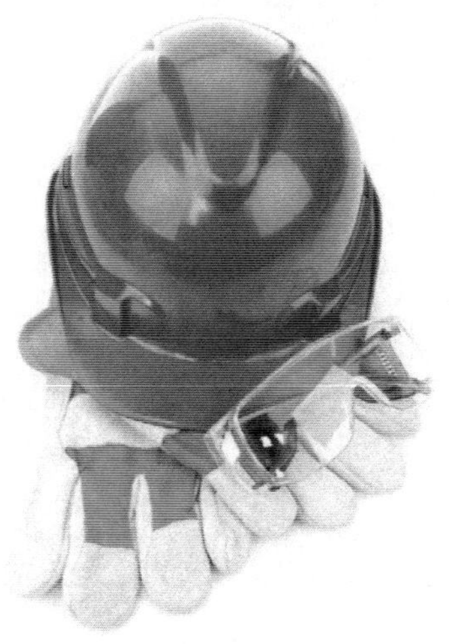

Tool 3 – Building in Sustainability - BONUS

- Developed an effectiveness algorithm on addressing causal factors based on scores for sustainability
- Can determine the cost of a CAR and its sustainability score
- Then compare the cost to the degree of risk reduction achieved with different CARs

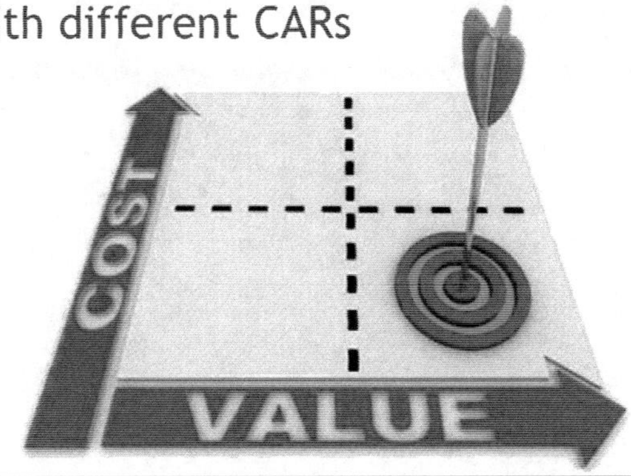

Sample Sustainable Recommendation

Causal Factor #1: Pin hole developed in pipeline

Root Cause	Basis for Specific Root Cause	Evidence	CA #	Corrective Action Recommendation
Equipment Difficulty – Preventive/Predictive Mainte[illegible] PM NI[illegible] for Equipment	Was a malfunction caused by a lack of preventive or predictive maintenance [illegible] Based Inspection (RBI) Pro[illegible] [illegible] subsea [illegible] less tha[illegible]	Interviews; company policy and [illegible]	1-1	Create an inspection and preventive/predictive maintenan[illegible] provide for [illegible] 16" as for those of 16" and larger.

Personnel trained in control

Audit results analyzed for root causes & actioned

Audits of the control

Which is more Effective?

Training for new personnel

Competency evaluation

Control conforms to best practice

Results reported to mgmt.

Audits of training compliance

Results reported to mgmt.

Tool 4 – Validating CARs

Would you validate my parking?

Good spacing on each side; pulled all the way in; could improve angle to sidewalk

Review in the field:

- Are validation reviews done six and 18 months after implementation?
- Were all parts of the CAR originally implemented?
- Are sustainability elements in place and functioning?
- Are CAR measurements communicated to management?
- Have there been further occurrences of the CF or RC?
- Are validation results communicated to management?

17 aramco

Summary

- SMART is only first part of developing good CARs
- Include QA hold points
- Connect the dots
- Include sustainability in your CARs
- Validate CARs in the field 6-18 months later

18 aramco

Saudi Aramco: Public

Questions

Larry Perkinson, CSP, ARM
perkinli@aramco.com
+966 13 873-3765

Management of Change - Critical Success Factors

- Conference: CCPS Middle East Conference on Process Safety
- Year: 2017
- Time:
  Wednesday, October 11, 2017 - 11:40am-12:00pm

Historic incidents like Flixborough (1974) have well emphasized the importance of Management of Change leading to the development of this as a standard part of the Process Safety Management Standard. the challenge is, how can we ensure that MOC is always effectively implemented? This presentation examines and identifies Critical Success Factors based on review of many Industry Best Practices that ensure effective MOC implementation at a Facility. These Critical Success Factors include scope of Change, effective Recognition of Change, quality MOC Review Process, Role play by MOC Committee, Change Workflow, Monitoring, KPIs, Software, Human Factors & Training. the Papers Also Identified Common Mistakes Made in the Development of MOC Program

Agenda

- Introduction
- What is Bogging?
- Signs of Bogging
- Causes of Bogging
- Bapco Bogging Protocol
- Case Study: 3CDU Fired Heater
- Concluding Remarks

Introduction

Bapco

- Bapco operates a 265 MBPD crude capacity refinery, started up in 1936
- At Bapco we have 45 fired heaters:
  - Various configurations
  - Different burner technologies
  - Ages ranging from 5 to 80 years
- Extensive operating experience with fired heaters and their optimisation and troubleshooting
- Full time Subject Matter Expert (SME) for fired heaters:
  - Assists operations departments in managing and improving energy and safety performance of fired heaters

Introduction (cont'd)

Bapco

- As part of the continuous improvement of our operations, Bapco implemented the Bogging Protocol:
  - Instructions to operators on how to detect fuel-rich conditions in fired equipment and take corrective action to prevent explosion

What is Bogging?

- Bogging is the operation of a fired heater with a fuel-rich atmosphere in the firebox
- This operating condition is known by many names e.g. "stalling", "smothering"
- It can be extremely dangerous
- It can quickly escalate into an explosion

What is Bogging? (cont'd)

stack flue gas colour
during bogging conditions

heater explosion following bogging

Signs of Bogging

Bapco

- Low O_2 and high CO/combustibles (incomplete combustion)
- Draft pulsation or "huffing" of the fired equipment
- Strong vibration
- Black smoke coming out of the stack
- Reduction in process fluid outlet temperature with constant or increasing fuel gas flow
- Strong fuel gas smell in the vicinity of the heater

Causes of Bogging

Bapco

Bogging is caused by insufficient combustion air or excess fuel gas, which results in incomplete combustion and accumulation of unburned hydrocarbons

- Inadequate opening of burner air registers and/or stack damper
- Induced and/or forced draft fan failure
- Air flow meter malfunction
- Fuel gas control valve malfunction
- Insufficient combustion air supply during rapid increases in heat demand e.g. during start-up, feed cut-in, addition of cold process fluid

Bapco Bogging Protocol

- Developed to prevent the escalation of bogging incidents
- It establishes a short series of steps to arrest the bogging and safely return the heater to normal operation
- The implementation includes training for the operators to prevent, identify and correct a bogging scenario

Bapco Bogging Protocol (cont'd)

For the purpose of our protocol fired equipment is considered bogged if:

- Excess oxygen (O_2) is below the low O_2 alarm or below 1.0 vol%, whichever is higher, and
- Carbon monoxide (CO)/combustibles are above upper limit of the analyser range or at 1500 vppm, whichever is lower.

Bapco Bogging Protocol (cont'd)

Special considerations:

- Fired equipment not provided with CO/combustibles analysers will rely on excess O_2 readings only
- The fired equipment must be assumed to be in a bogged condition if:
  - O_2 analyser fails and the CO/combustibles level is higher than the high alarm value
  - CO/combustibles analyser fails and the O_2 level is lower than the low O_2 alarm value

Bapco Bogging Protocol (cont'd)

Special considerations:

- Fired equipment not provided with analysers will be considered bogged whenever:
  - The coil outlet temperature (COT) decreases and the fuel gas flow rate increases

Bapco Bogging Protocol (cont'd)

بابكو Bapco

Bogging alerts configured in the DCS with appropriate alarms for each fired heater based on the excess O_2 and CO values defined on this protocol.

When the bogging alert is activated the Control Operator must:

- Place and hold the fuel gas control valve in manual mode
- Notify field crew of the situation and instruct them to evacuate all personnel within the vicinity of the heater
- Slowly reduce fuel gas flow until the all clear is given

Bapco Bogging Protocol (cont'd)

بابكو Bapco

After the bogging alert is cleared:

- Additional air can be added by opening the stack damper or air registers
- The fuel gas flow can then be increased, if required, to return the heater to normal conditions

Bapco Bogging Protocol (cont'd)

WARNINGS

- **<u>NEVER</u>** add air to a bogged fired heater because this could lead to an explosion
- If there is any indication that flameout has occurred, immediately evacuate the area and activate the emergency shutdown system

Case Study: 3CDU Fired Heater

- Fired heater located in No. 3 Crude Distillation Unit (3CDU)
- Designed to process 10 MBPD of crude
- Vertical cylindrical, natural draft, four burners firing upwards
- Not equipped with flue gas analysers
- Manually operated stack damper
- Burner air registers do not have a locking device to maintain the desired opening

Case Study: 3CDU Fired Heater (cont'd)

باپكو Bapco

Description of the incident:

- Reduction in heater outlet temperature and increase in fuel gas pressure and flow was observed
- Fuel gas control valve was switched to manual mode to reduce flow
- Strong vibration and black smoke from stack was observed
- Implemented actions were not enough to control heater firing
- The heater had to be shut down

Case Study: 3CDU Fired Heater (cont'd)

باپكو Bapco

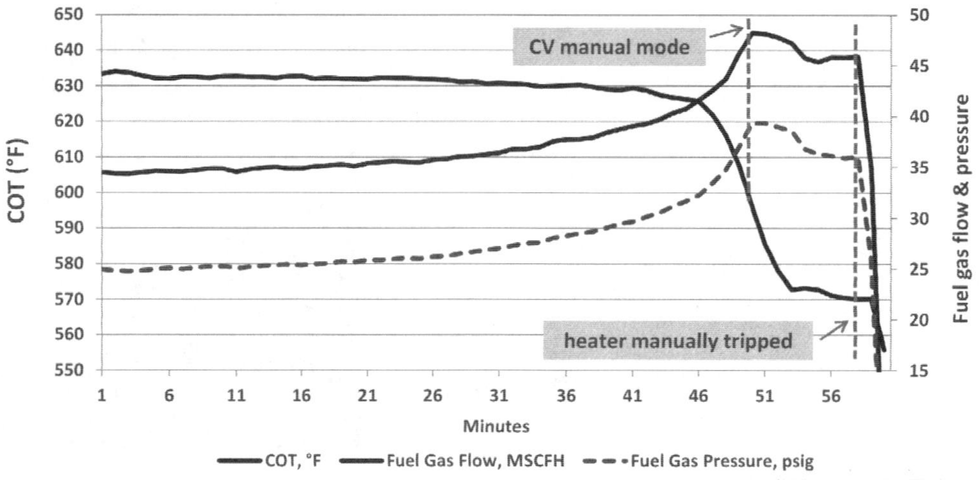

Case Study: 3CDU Fired Heater (cont'd)

Bapco

Observations:

- Coil outlet temperature (COT) dropped by 60 degF in 15 minutes
- Fuel gas pressure increased by 10 psi in 10 minutes
- Stack damper was seized at 50% opening
- Burner air registers were found almost closed
- Air registers were not provided with a locking device

Case Study: 3CDU Fired Heater (cont'd)

Bapco

Lessons learned:

- Bogging can develop quickly and pose a risk to personnel and assets
- Trends leading to bogging can be used for early detection
- Stack dampers must be operable at all times
- Burner air registers must have a lock-in-position device
- Operators must be trained on the prevention, detection and correction of bogging

Concluding Remarks

Bapco

- Bogging is an operating condition that poses a risk to personal and process safety
- Bapco is implementing a refinery-wide protocol to detect and correct bogging in fired heaters, boilers and incinerators
- The detection of bogging can be based on flue gas excess oxygen and combustibles readings or on coil outlet temperature and fuel gas flow
- Inadequate corrective actions can lead to an explosion, hence, operator training is a key success factor

Bapco

Thank You.

بابكو Bapco
دوماً نرتقي
Empowering Progress

Bapco

A REAL WORLD EXAMPLE OF MONITORING PROCESS SAFETY METRICS IN A REFINERY OVER FIVE YEARS

Bapco Operational Excellence Team

08 Month 2017

Bapco

BIO

- Dr Mark Hodgkinson-PhD (chemistry), BSc (Hon) Chemistry
- Dr Mark Hodgkinson holds is currently the OEMS Project Manager for Bapco. He holds a PhD and BSc in Chemistry and Chemometrics. He has over 25 years' experience in the Oil and Gas Field and also 15 years in Applied Microbiology fields such as water treatment.
- His professional interests include: Operational Excellence, Operational Intelligence, System Design , Change management, Business planning and Multivariate processes monitoring
- His leisure interests include Historical Board Gaming and Golf.
- OE Team ....+++++++

OUR BEGINNINGS 2010

EARLY GENERATION OF METRICS

- Bapco implemented an Operational Excellence Management System over the top of all our Management systems including Process Safety Management .
- Each Process has Leading and Lagging indicators

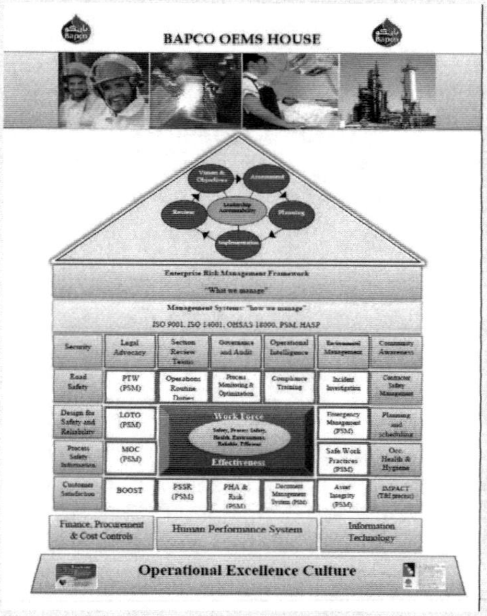

بابكو
Bapco

Note PSM Processes

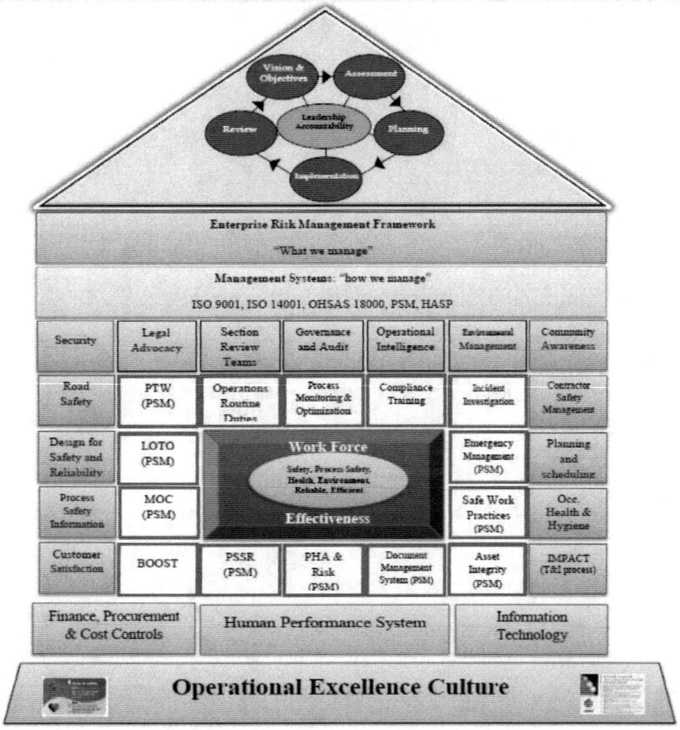

بابكو
Bapco

Metric Attributes

- All metrics are stated in the peak process documents (Process Descriptions) . We had leading and lagging indicators before API 754.
- An example of one (1) of the five (5) Leading indicators for this process

What is measured?	Leading: % Safety Critical Compliance checks for Relief Valves tests, Emergency Shutdown Device (ESD) tests and Over speed Trip Devices Tests (OTT). (PSV/RV, ESD Loops, Turbine trip tests)
Why is it measured?	Predicts this OEMS process performance by assessment of compliance.
Who measures it?	PMD – Maintenance Planning Section.
How is it measured?	Count of the actual checks and inspections performed against those that were scheduled for the month. Any re-schedule due to units unavailability counts as a miss.
Reporting format?	Tabular report to EHS Committee and Area Review and Monthly Tabular report via Operational Excellence Maintenance Measure (OEMM) by PMD Planning Section / XHQ.

METRIC ATTRIBUTES

- We used home built Spreadsheets to present summary results and detailed plots. All of this was very manual and ugly!
- Details of exactly what metrics to follow.

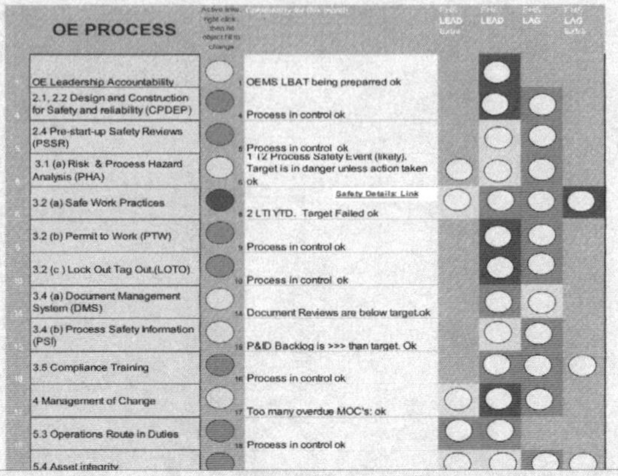

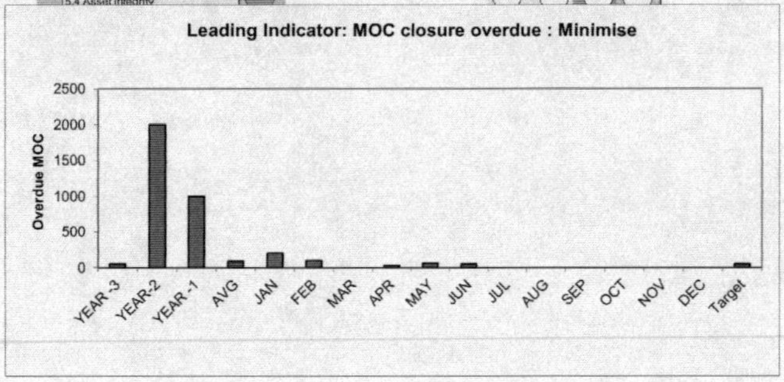

2013 ENHANCEMENTS

Early Generation of Metrics

- Bapco implemented an Operational Intelligence Platform (Siemen XHQ) with the KPI Solution in built
- We realized that we needed "health or Risk" metrics as well as Leading and Lagging indicators (Hopkins 2007)

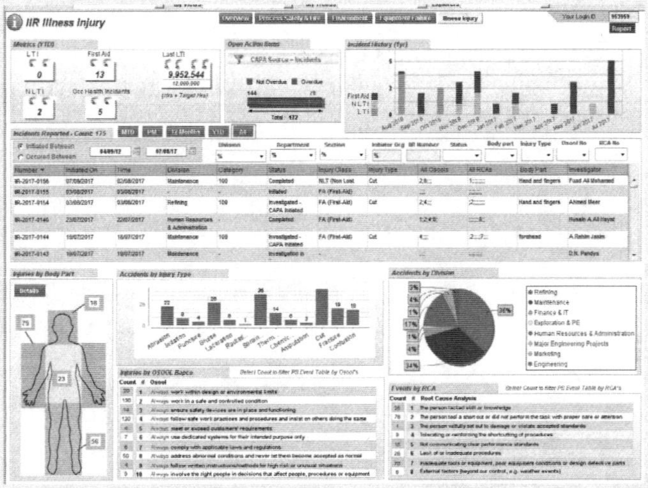

Metric Attributes

- The Lagging Indicator for the previous Leading Indicator.

What is measured?	Lagging; All Loss of Containment (API 754 T1, T2). ***This KPI measures the number of days between LOC events***
Why is it measured?	Measures the performance of this process.
Who measures it?	QA Specialist
How is it measured?	All Loss of Containments in the incident Investigation Database..
Reporting format?	Tabular report to EHS Committee and Area Review. / *XHQ*

THE CONCEPT (HOPKINS 2007)

Lead
- Leading Indicator
- Prediction from compliance, conformance, activities

Health
- Health Indicator
- The results of the Leading indicator activities

Lag
- Lagging Indicator
- Process Failure

EXAMPLE "HEALTH" METRICS

What is measured?	Relief and Safety Valves found to be failed to danger (PSVs that failed their pop test) API 754 T3)
Why is it measured?	Gives an indication of the risk level this process must control.
Who measures it?	PMD/MSD but reported by Reliability Engineering Dept.
How is it measured?	Number of Safety Device checks that find that the device may not have worked as required.
Reporting format?	Tabular report to EHS Committee, Area Review and XHQ.

Bapco

Early Generation of Metrics

- KPI Functionality included
- A wrap up to an index
- RAG
- Bar Charts

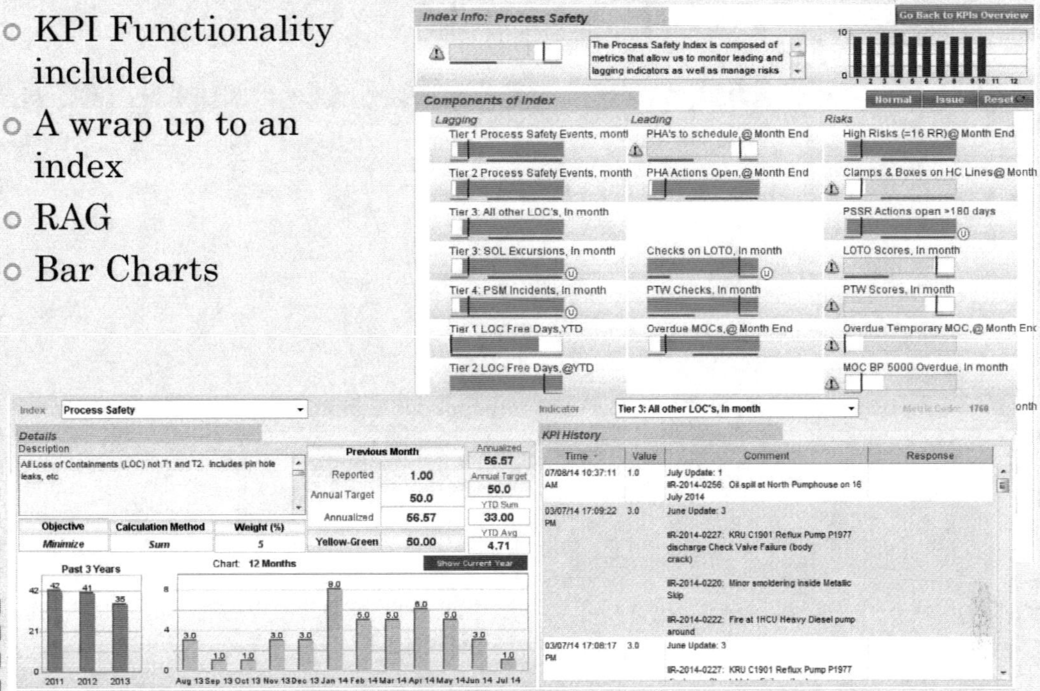

Bapco

2017 Enhancements

VIZION OE/PSM

- State of the art, (best in class) KPI/KOP System.

Input Screen

Vizion OE — Hello, BAPCODOM\xhqadmin

Performance Indicators

+ Add

Name	Type	Category	Impact		Order
Tier 1 Process Safety Events	EHSC PSM	General	1	V T	1
Tier 2 Process Safety Events	EHSC PSM	General	2	V T	3
% Relief Valves Failures	EHSC PSM	General	3	V T	18
ESD Checks that Failed %	EHSC PSM	General	3	V T	19
Add missing values her LOCs	EHSC PSM	General	3	V T	5
Tier 3 SOL excursions	EHSC PSM	General	3	V T	5
All Inspections deferred YTD	EHSC PSM	General	4	V T	23
Clamps & Boxes on Hydrocarbon Lines	EHSC PSM	General	4	V T	21
Days exceeding alarm limits in month	EHSC PSM	General	4	V T	26
High risks (RR=16) at month end	EHSC PSM	General	4	V T	12
LOTO Checks conducted	EHSC PSM	General	4	V T	15
LOTO Score %	EHSC PSM	General	4	V T	16
Open Actions in Proces Safety or Fire	EHSC PSM	General	4	V T	20
Overdue MOCs	EHSC PSM	General	4	V T	7
Overdue Temporary MOCs	EHSC PSM	General	4	V T	8
PHA Actions Open	EHSC PSM	General	4	V T	11
PHAs to schedule YTD	EHSC PSM	General	4	V T	10

VIZION OE/PSM

- State of the art, (best in class) KPI/KOP System.

Explorer

Vizion OE — Hello, BAPCODOM\xhq

Indicators Explorer

69	All
1	EHSC CHES
4	EHSC Emergency
11	EHSC Environment
3	EHSC Health
2	EHSC Hygiene
2	EHSC Leadership
14	EHSC Personal Safety
26	EHSC PSM
6	EHSC Road Safety

Status	Name	Projection	Impact
■	Tier 1 Process Safety Events	1.714	1
■	Tier 2 Process Safety Events	5.143	2
■	Tier 3 SOL excursions	6.857	3
■	Tier 3 all other LOCs	29.14	3
■	ESD Checks that Failed %	0	3
■	% Relief Valves Failures	0.8143	3
■	PHAs to schedule YTD	10.29	4
■	Safety critical checks %	100	4
■	LOTO Checks conducted	60.33	4
■	High risks (RR=16) at month end	0	4
■	Overdue Temporary MOCs	0.5	4
■	PSSR Actions Open > 180 days	0	4
■	Tier 4 Process Safety excursions	0	4
■	Clamps & Boxes on Hydrocarbon Lines	54.67	4
■	Days exceeding alarm limits in month	3.017	4
■	Tier 1 LOC free days	82	4
■	LOTO Score %	96.7	4
■	PTW Checks conducted	231.2	4
■	Unexpected Corrosion Events YTD	22	4
■	All Inspections deferred YTD	10	4

VIZION OE/PSM

- State of the art, (best in class) KPI/KOP System.

Flexible Plots

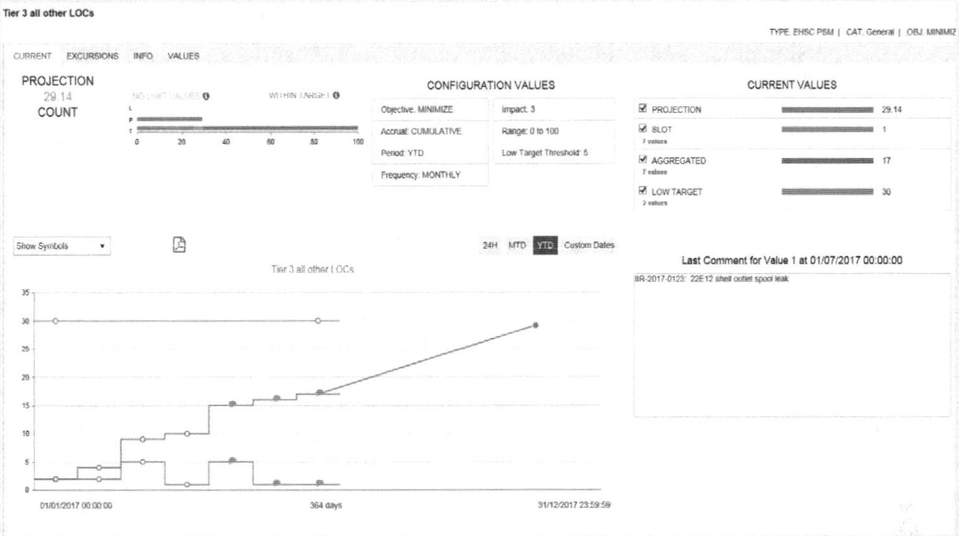

PSM METRICS MAP

PSM METRICS

PSM Process	Leading	Health/Risk	Lagging
Process Safety Information.	Drawings and data sheet up date conformance		Incidents where a root cause includes PSI found to be missing or wrong.
Process Hazard Analysis.	PHA to Schedule Open PHA Actions Overdue PHA actions	PHA Actions > Risk Rank X	Incident investigations that show the root cause includes PHA was not done or action item from PHA was not implemented in the specified time period.
Operating Procedures	Ops and Mtce Document Reviews Over due	Ops and Mtce Documents changed per year	Incident investigations that show the root cause includes poor or missing document

PSM METRICS

PSM Process	Leading	Health/Risk	Lagging
Training	Mandatory Training conformance.		Incidents where a root cause includes Training Issues
Contractors Safety Management	% Contractor Safety Meetings to schedule	Number of safety violations by contractor	Contractors Statistics (LTI, NLTI, FA).
Pre-Start Up Safety Review	PSSR Actions overdue (180 days)		Lagging: PSSR faults -Post Start up PSSR incidents captured under Tier 4 : PSM incidents
Mechanical/Asset Integrity	Compliance of Statutory Checks /Inspections on equipment and safety critical devices LOTO, PTW checks	The result of the checks. Safe Operating Limit Excursions. Boxes/Clamps on H/C lines. Unexpected Corrosion events. Alarms levels.	Loss of Containments T1,T2, T3

PSM Metrics

PSM Process	Leading	Health/Risk	Lagging
Safe Work Practice PTW, Isolation, Confined Space, Stop Work, Excavation. Others…..	Leading: Conformance to standard for working at heights, Excavation and Falling objects. BBS.	% Behavioural showing PPE Compliance PTW, LOTO, Safety Audit/checks/Inspecti on findings.	Lagging : Injury Frequency Rate
Management of Change.	MOC's overdue Temporary MOC's O/D	MOC Audit findings	Incidents where MOC failure was a Root Cause.
Incident Investigation	Overdue Incidents Investigations Near Miss to Incident ratio.	Incident Investigation CAPA close out rate	Repeat Incidents
Emergency Planning and Response	Emergency Exercises Fire Drills	Fire Equipment Readiness	Fires API 754 T1, T2
Compliance Audit Governance and Audit	Audits to schedule	Audit findings CAPA close out rate	Tier 1 and 2 HSE Incidents.

A worked Example

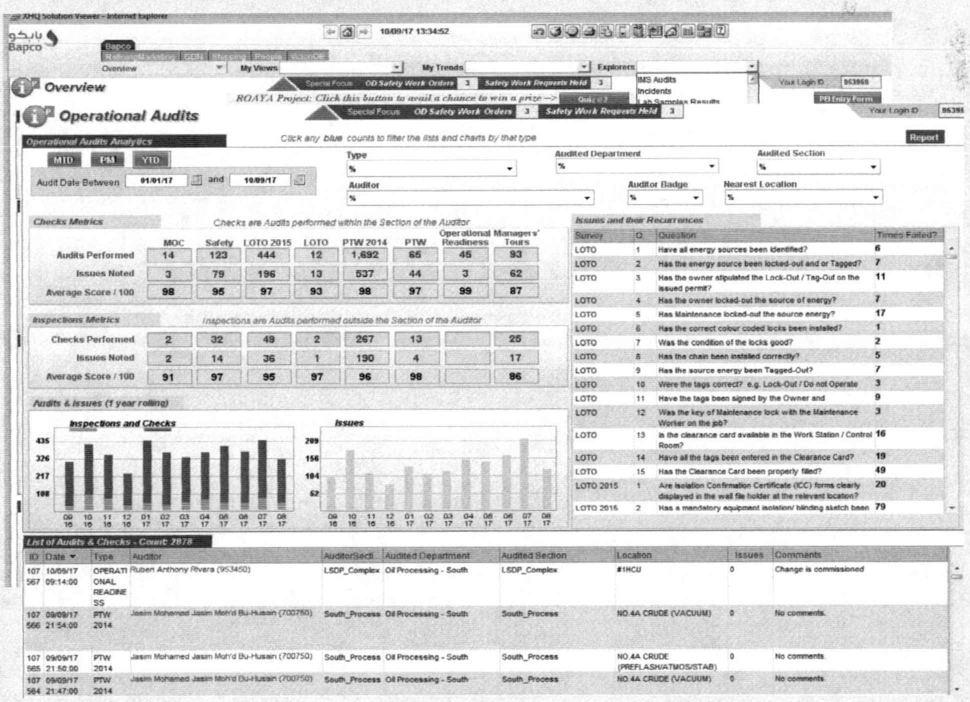

بابكو
Bapco

A WORKED EXAMPLE

Audit Information

Date / Time	Survey Department	Survey Location
8/22/2017 11:41:00 PM	**Oil Processing - North**	**NO.1 VDU**
Survey Type	Survey Section	Org Code
PTW 2014	**North_Process**	**03RN11**

Additional Information

Receiving Authority:	874215
Description of Work:	inspect burner of heater 2F1
Safe Work Permit Number	394340
Associated Certificates/ Forms	JSA:ICC:

Auditor

Auditor	Department	Org Code
Ahmed Hasan Ahmed Ashoor	**Oil Processing - North**	**03RN11**
Badge	Section	
708795	**North_Process**	

	Question	Answer
1	Is the scope of work clearly identified?	YES
2	Are necessary risk assessments available for review?	YES
3	Are identified hazards listed on the permit to work, JSA or RAMS?	YES
4	Are appropriate precautions listed on the permit to work (including clearly specified isolations)?	YES
5	Is the operational time limit of the permit clear? (are extensions properly authorised?)	YES
6	Are certificates completed properly and appended to the permit?	YES
7	Are other area or system activities cross-referenced correctly via the permit?	N/A
8	Are copies of permits, certificates and attachments legible?	YES
9	Are signatures and initials traceable and legible?	YES
10	Are copies of permits and certificates posted at the correct locations?	YES
11	Are attachments, drawings etc. held at the correct locations?	N/A
12	Are Users briefed on the permit to work, and have they acknowledged understanding of requrements?	YES
13	Do people know what to do in the event of emergency?	YES
14	Are isolations appropriate for the task, clearly specified on the isolation certificate, and correctly implemented?	YES
15	Are common isolations cross-referenced?	N/A
16	Are the right people aware of isolated equipment?	YES
17	Is the Owner Dept. aware of the work?	YES
18	Is the work being carried out in conformance with the permit and associated certificates?	YES
19	Are control measures and personal protective equipment appropriate for the task?	YES

sues and their Recurrences

urvey	Q.	Question ▲	Times Failed?
		RAMS?	
TW 2014	14	Are isolations appropriate for the task, clearly specified on the isolation certificate, and correctly implemented?	
TW 2014	2	Are necessary risk assessments available for review?	
TW 2014	7	Are other area or system activities cross-referenced correctly via the permit?	
TW 2014	22	Are safety compliance spot checks being carried out and recorded on the original copy of the SWP?	1
TW 2014	9	Are signatures and initials traceable and legible?	2
TW 2014	16	Are the right people aware of isolated equipment?	2
TW 2014	20	Are tools and equipment suitable and in good condition?	16
TW 2014	13	Do people know what to do in the event of emergency?	
TW 2014	23	Have the SWP and associated Certificate forms been cancelled properly?	3
TW 2014	17	Is the Owner Dept. aware of the work?	17
TW 2014	5	Is the operational time limit of the permit clear? (are extensions properly authorised?)	8
TW 2014	1	Is the scope of work clearly identified?	4
TW 2014	18	Is the work being carried out in conformance with the permit and associated certificates?	4

بابكو
Bapco

THANK YOU

ابن سينا
ibn sina

USE PLANT WALKTHROUGHS TO ENHANCE PROCESS SAFETY RISK DISCOVERY

October, 2017

By
Mohammed Saleh Al-Kanfari
Sami S. Al-MAIYAD

A SABIC and CTE Joint Venture

ابن سينا
ibn sina

CONTENTS

No. 1

- INTRODUCTION

PROCESS SAFETY AND PERSONAL SAFETY

PROCESS SAFETY VS PERSONAL SAFETY

ابن سينا
ibn sina

WHAT IS PROCESS SAFETY TO YOU?

Personal Safety

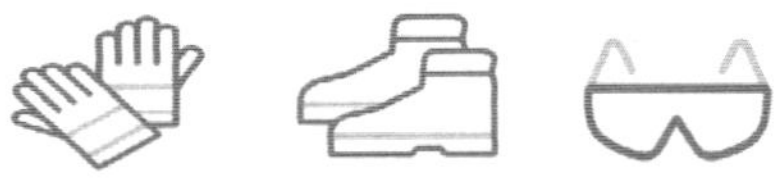

Personal Safety Protecting against personal injuries
Slips, trips, falls, etc.

Process Safety

Process Safety is a blend of engineering and management skills focused on preventing catastrophic accidents, particularly explosions, fires, and toxic releases

No. 3

ابن سينا
ibn sina

PERSONAL SAFETY VS PROCESS SAFETY

Personal **Safety**	Process **Safety**
Frequent	
Higher frequency occurrences	Lower frequency occurrences
What does it prevent?	
Health and safety issue related to trips, slips, falls, exposure to chemicals, noise, etc resulting hart at an individual level.	Unintentional releases of chemicals and energy leading to major event with big consequences (toxic releases, fires, explosions, pollution)
Who does it protect?	
Workers	Workers, public, business, environment.
What does it try to act on?	
Individuals' behavior, Management systems.	System design (Processes, Equipment), Management systems, Individuals behavior.
How Difficult is it to grasp?	
Easier to understand (not mean easy to implemented).	Requires complex engineering and multidisciplinary knowledge.

No. 4

- PURPOSE OF PS WALKTHROUGH PROGRAM

ابن سينا
ibn sina

THE PURPOSE OF PS WALKTHROUGH PROGRAM

- To identify hidden process safety (PS) risks.
- To ensure an effective system exists to manage PS Walkthrough which will help in ensuring strength of the protection barriers or through procedures.
- To establish guidelines for carrying out PS Walkthrough by using pre-defined checklists.
- To assign responsibilities for effective implementation the system.
- To do analysis of the findings from PS Walkthrough and investigate common causes and recommendation to ensure the findings will not be repeated.

No. 6

ابن سينا
ibn sina

THE PURPOSE OF PS WALKTHROUGH PROGRAM

Scope

This program defines requirements for PS Walkthrough activity for all the process plants, storage and utilities

Expected results

- Zero overdue of PS Walkthrough rounds.
- Zero overdue of PS Walkthrough finding actions.

No. 7

- 1. FREQUENCY & TEAM COMPOSITION OF PS WALKTHROUGH ROUNDS
  2. RESPONSIBILITIES

ابن سينا
ibn sina

FREQUENCY & TEAM COMPOSITION OF PS WALKTHROUGH ROUNDS

Weekly

- One PS Walkthrough Round should be conducted Weekly .
- Each shift will be responsible to conduct 4 rounds per month.
- The team members should be as following:
  - Plant specialist or shift supervisor.
  - Minimum two operator.

Quarterly

- Executive shall have a quarterly rounds to cover Process Safety Walkthrough checklists as following:
  - 1 rounds by GMT and one by GMO.
  - 1 round by Vise President, and
  - 1 round by President.
- The team members should be as following:
  - Operation/ TSD/ EHSS/ maintenance Sr. Mangers.
  - Process safety engineer.
  - Plant specialist/shift supervisor or Production Engineer.

Monthly

- One PS Walkthrough Round should be conducted monthly.
- Operation Sr. Manger Shall lead at least 6 PS Walkthrough rounds per year and may assign 6 by Operation manger.
- The team members should be as following:
  - Manager (Inspection/process/rotating/ safety).
  - Plant specialist or shift supervisor or Production Engineer.
  - Process safety engineer
  - Inspection Engineer
  - Safety Engineer or specialist.
  - Electrical/ Instrument engineers (as required).
  - Minimum one operator.

No. 9

ابن سينا
ibn sina

RESPONSIBILITIES

1 PS Focal Point (PS Engineer) → 2 Operation Manager → 3 Operation Sr.Manager → 4 Rounds Leader

- Develops and issue annual PS Walkthrough schedule for each plant
- To ensure that PS Walkthrough are conducted as per schedule.
- Provide PS Walkthrough status to EHSS Steering committee on monthly basis
- Carry out the analysis of all the findings on annual basis.
- Shall conduct PS Walkthrough awareness when needed.
- Shall ensure that PS Walkthrough process is effective and improved continuously.

ابن سينا
ibn sina

RESPONSIBILITIES

1 PS Focal Point (PS Engineer) → **2 Operation Manager** → 3 Operation Sr.Manager → 4 Rounds Leader

- Shall ensure that PS Walkthrough are conducted for his unit as per schedule.
- Shall track unit level actions of findings of all PS Walkthrough and assign those actions to concerned/required disciplines.
- Shall provide status of actions to PS Walkthrough Focal Point on monthly basis.

ابن سينا
ibn sina

RESPONSIBILITIES

1 PS Focal Point (PS Engineer) › 2 Operation Manager › **3 Operation Sr. Managers** › 4 Rounds Leader

- Ensure effective implementation of the system.
- Lead the related PS Walkthrough for 12 rounds per year.
- May assign 6 PS Walkthrough rounds per year out of 12 to be led by operation manager.

ابن سينا
ibn sina

RESPONSIBILITIES

1 PS Focal Point (PS Engineer) › 2 Operation Manager › 3 Operation Sr.Managers › **4 Rounds Leader**

- Prepare PS Walkthrough by:
  - Review previous PS Walkthrough check list findings and Record PS Walkthrough team findings on master hard copy
  - Issue draft soft copy of filled PS Walkthrough checklist for comments and corrections to the team.
  - Issue soft copy of filled PS Walkthrough checklist to Unit Operations Manager and PS focal point for tracking purpose.

ابن سينا
ibn sina

EXAMPLE PROGRAM SCHEDULE

Action	Target	BY	Remarks
Create PS walkthrough check list	• July 2016	• PS Engineers	
Issue PS walkthrough procedure	• July 2016	• PS Engineers	
Conduct PS Walkthrough awareness to concern members.	• July/August	• PS Engineers	
Conduct Pilot PS walkthrough	• August 2016	• PS Engineers + Team	
Lunching PS walkthrough program and conduct first PS walkthrough	• End of August	• Management	

No. 14

• PS WALKTHROUGH CHECKLIST

PS WALKTHROUGH CHECKLIST

- The PS walkthrough check list divided into sections for key topics (examples following):

1. SAFETY, HEALTH AND FIRE PROTECTION	**2. SAFETY & RELIEF FACILITIES**	**3. PIPING, VALVES, VESSELS AND FURNACES**
4. PUMPS / COMPRESSORS	**5. WASTER STREAM, ENVOIRONMENT, UTILITY SYSTEMS, TANKS**	**6. HOSES**

No. 16

EXAMPLE PS WALKTHROUGH CHECKLIST

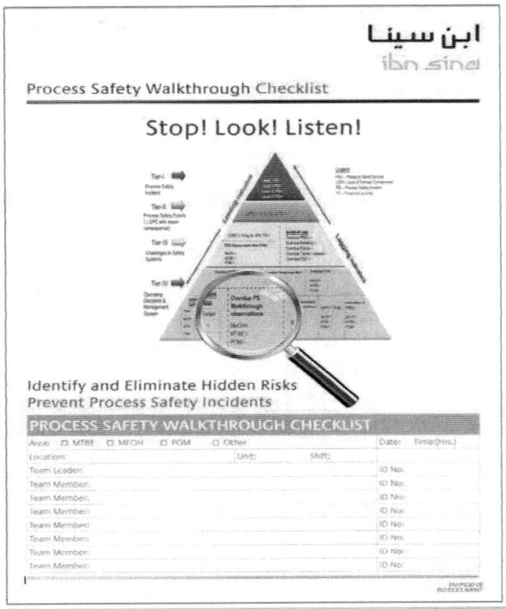

ITEMS	CHECK ITEMS	REMARKS
A	SAFETY, HEALTH AND FIRE PROTECTION	
1	Are process sight glasses, flow indicators, gauges, etc. in good condition and reliable	
2	Do sighs adequately identify work are hazards and equipment provided?	
3	Is unobstructed access to safety and fire protection equipment provided?	
B	SAFETY & RELIEF FACILITIES	
1	Is there signs of corrosion in the body of the PSV?	
2	Is there sign of passing PSV?	
3	Is safety valve inlet and outlet piping properly supported?	
4	Is atmospheric safety valve provided with weep (drain) hole at bottom?	
5	Have you observed PSV chattering (Rapid opening and closing of PRV)?	
6	Are all flare tips pilots ignited?	
7	Is flare base water level within normal range?	
C	PIPING, VALVES, VESSELS AND FURNACES	
1	Have you noticed crack or leak in piping?	
2	Have you noticed Damage insulation?	
3	Have you noticed any hot spot in piping?	
4	Is there missing or damaged plug or cap in pipe connections?	
5	Is there pool support piping?	
6	Is there signs of vent pluggage?	
7	Is there missing or damaged car seal?	
8	Is there proper support to the temporary clamp?	
9	Is there sign of leak from temporary clamp?	
10	Are vents and drains located such that they do not create personnel hazards?	
11	Are sample points property configured for safe sampling?	
12	Do you notice abnormal flame shape?	
13	LABELLING: Have lines been clearly labelled, including flow arrows (special attention for battery limit isolation valve)? Have vessels been properly labelled?	
D	PUMPS / COMPRESSORS	
1	Is there liquid around the pump? What type of liquid?	
2	Is there abnormal sound from pump?	
3	Is there enough fluid in the reservoir (seal plot) and is it flowing to the pump?	
4	Do you notice hot motor surfaces that could be an ignition source?	
5	Is there evidence of pump seal leak? Is there enough level in the seal pot? Check pressure and flow indications.	
E	WASTER STREAM, ENVOIRONMENT, UTILITY SYSTEMS, TANKS & STORAGE	
1	Have you noticed any damage in secondary containment (dike)?	
2	Have dinking isolation valves been closed?	
3	Do you observe any sign of corrosion or leak from bottom of storage tank?	
4	Do you notice waste is properly segregated and labelled?	
5	Do you notice any smoke from flare or furnaces or boilers?	
F	HOSES	
1	Is there signs of corrosion or hose leak?	
2	Are the seals (gasket or O-rings) in good condition?	
3	Is the hose provided with valid inspection tag?	
4	Is there a hose improper support?	
5	Is proper type of hose used?	
6	Are there utility or process hoses without non return valve/	
7	Are there temporary hoses connected to process equipment w/o proper approval/temporary MOC (hoses used for wrong service)?	

No. 17

ابن سينا
ibn sina

PS WALKTHROUGH CHECKLIST

ITEMS	CHECK ITEMS	REMARKS
A	SAFETY, HEALTH AND FIRE PROTECTION	
1	Are process sight glasses, flow indicators, gauges, etc. in good condition and reliable	
2	Do sigs adequately identify work are hazards and equipment provided?	
3	Is unobstructed access to safety and fire protection equipment provided?	
B	SAFETY & RELIEF FACILITIES	
1	Is there signs of corrosion in the body of the PSV?	
2	Is there sign of passing PSV?	
3	Is safety valve inlet and outlet piping properly supported?	
4	Is atmospheric safety valve provided with weep (drain) hole at bottom?	
5	Have you observed PSV chattering (Rapid opening and closing of PRV)?	
6	Are all flare tips pilots ignited?	
7	Is flare base water level within normal range?	

No. 18

ابن سينا
ibn sina

PS WALKTHROUGH CHECKLIST

C	PIPING, VALVES, VESSELS AND FURNACES	
1	Have you noticed crack or leak in piping?	
2	Have you noticed Damage insulation?	
3	Have you noticed any hot spot in piping?	
4	Is there missing or damaged plug or cap in pipe connections?	
5	Is there pool support piping?	
6	Is there signs of vent pluggage?	
7	Is there missing or damaged car seal?	
8	Is there proper support to the temporary clamp?	
9	Is there sign of leak from temporary clamp?	
10	Are vents and drains located such that they do not create personnel hazards?	
11	Are sample points property configured for safe sampling?	
12	Do you notice abnormal flame shape?	
13	LABELLING: Have lines been clearly labelled, including flow arrows (special attention for battery limit isolation valve)? Have vessels been properly labelled?	
D	PUMPS / COMPRESSORS	
1	Is there liquid around the pump? What type of liquid?	
2	Is there abnormal sound from pump?	
3	Is there enough fluid in the reservoir (seal plot) and is it flowing to the pump?	
4	Do you notice hot motor surfaces that could be an ignition source?	
5	Is there evidence of pump seal leak? Is there enough level in the seal pot? Check pressure and flow indications.	

No. 19

ابن سينا
ibn sina

PS WALKTHROUGH CHECKLIST

E	WASTER STREAM, ENVOIRONMENT, UTILITY SYSTEMS, TANKS & STORAGE	
1	Have you noticed any damage in secondary containment (dike)?	
2	Have dinking isolation valves been closed?	
3	Do you observe any sign of corrosion or leak from bottom of storage tank?	
4	Do you notice waste is properly segregated and labelled?	
5	Do you notice any smoke from flare or furnaces or boilers?	
F	HOSES	
1	Is there signs of corrosion or hose leak?	
2	Are the seals (gasket or O-rings) in good condition?	
3	Is the hose provided with valid inspection tag?	
4	Is there a hose improper support?	
5	Is proper type of hose used?	
6	Are there utility or process hoses without non return valve/	
7	Are there temporary hoses connected to process equipment w/o proper approval/temporary MOC (hoses used for wrong service)?	

No. 20

- ANALYSIS OF THE FINDINGS

PROCESS SAFETY WALKTHROUGH EXAMPLE ANALYSIS

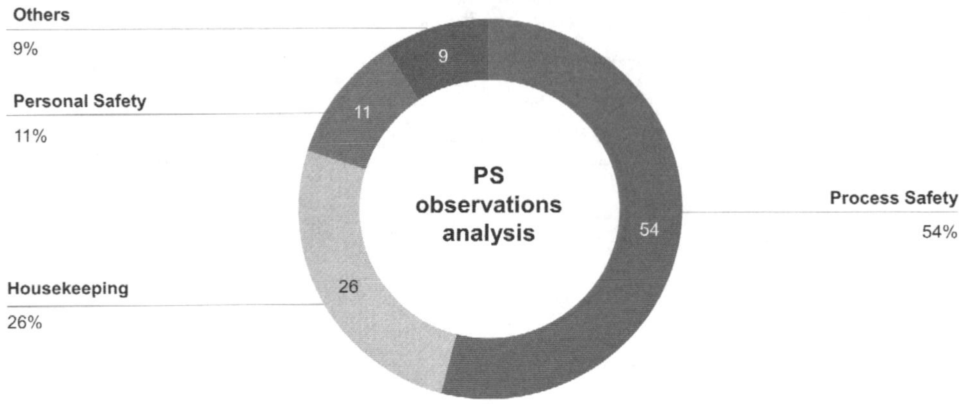

PROCESS SAFETY WALKTHROUGH EXAMPLE ANALYSIS

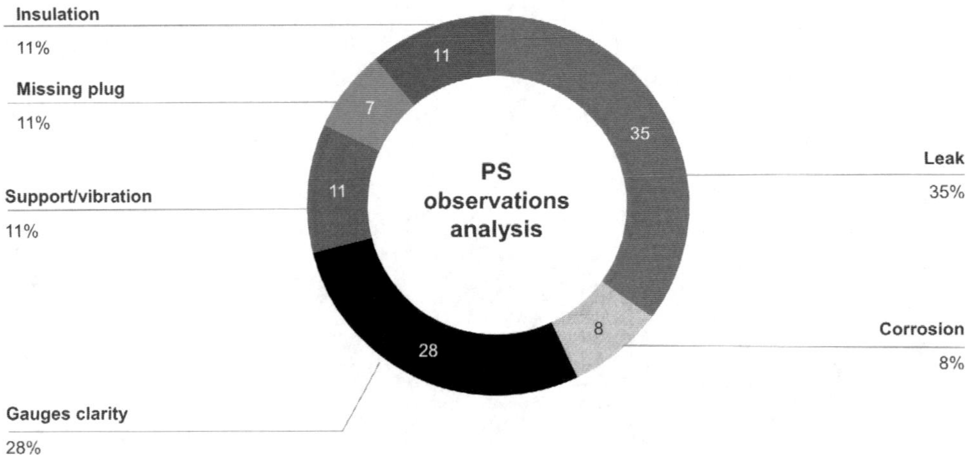

ابن سينا
ibn sina

EXAMPLE FINDINGS FROM PS WALKTHROUGH

PLANT #1

- Dusty sight glasses, flow indicators, gauges.
- Unreliable pressure gauges.
- PSV's have hole in the bottom of the discharge pipe plugged.
- Figure 8 blind for HE-xxx in the wrong position (open position instead of close position).

PLANT #2

- Equalizer valve of PV-xxx has pinhole leak.
- MeOH leak in the new clamp installed in HE-xxx.
- Observed CO leak on the isolation valve downstream of UV-xxx.
- PSV observed dripping
- Air House used for oil service.
- Cracks found in tank farm area outside and inside Dikes.

No. 24

ابن سينا
ibn sina

PICTURES

Lunching PS walkthrough program

Recognition by executive management for best PS finding

No. 25

ابن سينا
ibn sina

Thank You

No. 26

ابن سينا
ibn sina

NO WARRANTY

This information is provided solely for discussion purposes only. Ibn Sina MAKES NO WARRANTY, EXPRESS OR IMPLIED, INCLUDING MERCHANTABILITY AND NONINFRINGEMENT OF INTELLECTUAL PROPERTY, NOR ASSUMES ANY LIABILITY, DIRECT OR INDIRECT, WITH RESPECT TO THE PERFORMANCE, SUITABILITY OR FITNESS FOR INTENDED USE OR PURPOSE OF THIS INFORMATION IN ANY APPLICATION. Each user must determine the suitability of this information for the their particular use. No statement concerning a possible use of any product, service or design is intended, or should be construed, to grant any license under any patent or other intellectual property right.

No. 27

Process Safety Leading and Lagging KPI – "You Don't Improve What You Don't Measure"

- Conference: CCPS Middle East Conference on Process Safety
- Year: 2017
- Time:
  Wednesday, October 11, 2017 - 12:00pm-12:30pm

The paper will explain the need and necessity for having a process safety key performance Indicators, and explore the benefits and challenges of identifying , monitoring and controlling those indicators toward plant process safety. Examples of KPIs for process safety management systems will be explained and categorizing as leading and lagging KPIs , four tier approach will be demonstrated with recommended indicator for each tier . Additionally, the importance of selecting and balancing leading and lagging indicators effect on predicting and avoiding any incident. Moreover, share examples for automated KPI created at Saudkayan to monitor and control specific leading process safety KPI. Finally, list of recommended process safety metrics and compare it to OSHA requirements.

AUTHOR INDEX